Water Hammer

Problems and Solutions

B B Sharp

Reader in Civil Engineering
University of Melbourne

Edward Arnold

© B B Sharp 1981

First published in 1981
by Edward Arnold (Publishers) Ltd
41 Bedford Square, London WC1B 3DQ

British Library Cataloguing in Publication Data

Sharp, B B
  Water hammer.
  1. Water hammer
  I. Title
  621.8′672   TC174

ISBN 0 7131 3427 5

Typeset by Macmillan India Ltd, Bangalore

Preface

Water hammer is not normally given as an extensive instruction in engineering courses – it is one of those areas left for experts to exploit. There is nevertheless a growing awareness that water hammer should be the dominant design consideration when pipeline systems are being developed. Unfortunately it does require some study and familiarity with wave mechanics before one is able to tackle the problems that need solving.

There is a need for some direct answers to seemingly simple problems which will either convince the engineer that he should take the time to learn a great deal more about water hammer or that he should learn to respect the phenomenon for its importance and recognize the depth of design which is in reality required to analyse pipe systems of all kinds properly. Thus this text, whilst laying the groundwork for the understanding of the physics of water hammer in the simplest of terms, demonstrates the solution to a range of problems in most cases by two methods, which should point to the state of the art. The methods used are not exhaustive; mathematical rigour is the province of the mathematician.

The study of water hammer has been somewhat slowed down because computer programming techniques are essential for the solution of complex problems. This text includes quite simple computer programs which can be combined to form a basis for analysis.

The material used has evolved over a long association with the teaching of water hammer in civil engineering courses at undergraduate and postgraduate level in the University of Melbourne. Considerable consulting activity has led to the development of computer solutions which relate to real engineering needs. The initial interest, however, began many years ago when the author was associated with a very experienced and original design engineer, John S Gerny of the Engineering and Water Supply Department of South Australia.

The help of the University of Melbourne is gratefully acknowledged, in particular for the access to computer facilities, and, for the gradual evolvement of the material, the advice and patience of close colleagues and my wife are greatly appreciated.

Publication of this book has been assisted by a grant from the Committee on Research and Graduate Studies of the University of Melbourne.

B B Sharp
1980

Contents

Symbols

a_1, a_2, A	areas
A_g	area of gate
A_p	area of pipe
b	coefficient of velocity
C, C_0, C_a	wave speed (celerity)
C^-, C^+	characteristics
C_d	coefficient of discharge
D	diameter
e	base of natural logarithms
e	wall thickness
E	Young's elastic modulus, efficiency ratio
f, F	arbitrary wave functions, friction factor
g	acceleration (gravity), wave variable
GD^2	machine inertia (metric)
h, H	pressure head
H_{abs}	pressure head absolute
H_0	orifice pressure head loss
h_f	friction head loss
h_0	reference pressure head
I	second moment of area
i	$\sqrt{-1}$
k	friction constant, constant
K	liquid elastic modulus, constant
L	distance, length
M	constant (torque related), velocity ratio, constant
n	exponent
N	machine speed, velocity ratio, constant
p	constant, pressure head
p_0	atmospheric pressure
P_0	power
q, Q	discharge
r	ratio
R	hydraulic mean radius, ratio
S	slope

t	time
t_0	reference time
t_x	time at distance x
T	time, L/C_0 units, period
T_c	time of closure
T	machine torque
v, V	velocity
u	wave variable
v_0	reference velocity
v_r	residual velocity
V, V'	air vessel volume
WR^2	machine inertia (Imperial units)
x, x_1	distance
y, Y	pressure head
Z, z	vertical distance, impedance
α	ratio pump speeds, friction ratio, part of complex number
β	gate parameter, ratio pump torques, part of complex number
γ	pressure head, propagation constant
δ, Δ	increment
ε	strain
η	machine efficiency
θ	angle
λ	g/C_0
μ	Poisson's ratio
v	wave variable
π	3.14159
ρ	density
∂	partial differential
σ	surface tension, stress
Σ	sum
τ_0	shear stress
ϕ	wave variable, angle
ω	angular velocity

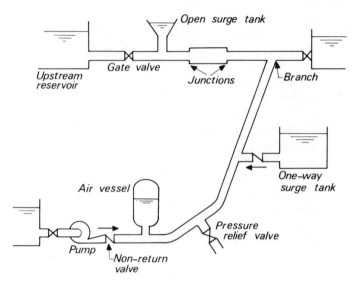

Open surge tank

Gate valve

Upstream
reservoir

Junctions

Branch

One-way
surge tank

Air vessel

Pressure
relief valve

Pump

Non-return
valve

1

Water hammer

The term water hammer itself is expressive and in most cases its presence and potential for damage is distinguished by the sound associated with it. There are many cases, however, where water hammer has not been heard but has caused pipeline failure and the circumstances and proof that water hammer was responsible have only been established after extensive analysis. It is fleeting and can prove extremely difficult to quantify in a field situation.

Water hammer is produced by a sudden change in the velocity of flow in a conduit, e.g.

(1) The stopping and starting of pumps
(2) The turning off of a valve at a wash basin – especially assisted by a spring loaded return mechanism
(3) The sudden flow demand by an automatic fire protection system – the antithesis of water hammer
(4) The fail safe protection systems which require rapid operation such as closure of a liquid fuel supply
(5) A resonance build-up due to a dynamically unstable component in a liquid line
(6) Mechanical failure of an item such as a valve – a casualty of age or wear from repeated use.

Water hammer became the subject of extensive publication after Allievi (1913) and contributors to the first Symposium on Water Hammer (1933) expounded a number of the fundamentals. More recently, the introduction of new materials such as plastics and asbestos cement and more sophisticated methods of system control and optimization have increased the necessity for an understanding of the subject.

Methods of analysis were originally carried out by numerical solutions of series of algebraic equations, and then graphical methods were evolved. Today, the modern computer allows problems of complex systems to be solved. However, like many technical areas, the prescription for satisfactory use is not so easily identified and the simple solution is not yet available. There are still areas of inadequate analysis and there is a need for a thorough grounding in the physics of water hammer so that those limitations can be appreciated and a suitable engineering solution found.

1

2 Water hammer problems and solutions

Water hammer is the propagation of energy as in the transmission of sound and is known from basic physics as a wave motion which is associated with the elastic deformation of the medium. The celerity of sound waves, C, is expressed as

$$C = \sqrt{\frac{K}{\rho}} \qquad (1.1)$$

in which K is the elastic modulus of the medium and ρ its density. The term celerity of a wave is used to differentiate between the velocity of a moving object and the velocity of a wave, the latter requiring only the unsteady motion of the object about a mean position or with respect to a mean motion. For example, sound waves at ground level pass through air at normal temperatures with a celerity of

$$C_a = \sqrt{\frac{118\,500}{1.293}} = 344 \text{ m s}^{-1} \quad \begin{array}{l}(1130 \text{ feet per second,} \\ 1250 \text{ km per hour})\end{array}$$

and the speed of the air molecules themselves is ordinarily of a much lower order than this.

This situation also exists in water hammer. Sudden changes in a pipe full of water may be propagated at a celerity of 1200 m s^{-1} and the local liquid velocities are not normally greater than 6 m s^{-1}. Since these speeds are, for the purposes of this text, never of comparable order it has become customary to use the expression 'velocity' of the water hammer waves since there can never be confusion between them and the local liquid velocities.

When water hammer is produced it is transmitted to all parts of the connected system, where it is modified by changes in the flow geometry and the conduit properties. A sudden change of 1 m s^{-1} can be accompanied by 100 m pressure head change. Valve controls and associated pipework, for example, should be designed so that changes do not result in the allowable structural stresses being exceeded. However, it would obviously be aggravating in the control of a system if, instead of a steady or slowly-changing flow, there were persistent, sudden, extreme changes of the pressure head and flow due to water hammer.

A pump at A in Fig. 1.1, when started, would produce a fluctuation of pressure head with time as shown in the lower diagram. The pressure head versus time trace seen by a stationary observer at A suggests water hammer waves. However, if the observer were travelling in the pipe with the pressure wave, at approximately 1200 m s^{-1}, he would be conscious only of the attenuation due to effects such as friction. Upon reaching some change in geometry, modifications due to transmission and reflection would require

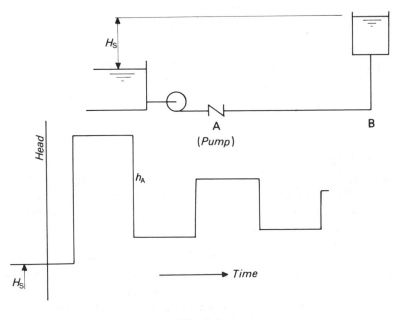

Fig. 1.1

him to divide his interest in different directions forward and backward in order not to lose track of subsequent events.

The orderly assembly of all the necessary data and an understanding of the properties of wave transmission and reflection are the first steps in the preparation of the analysis of any scheme. Once a system has been analysed for its (dynamic) water hammer response, the maximum and minimum pressures that might occur are then known and the system may be modified if necessary by the addition of some protective control measure and then re-analysed.

2

Basic equations, wave speed

2.1 Introduction

The analysis of water hammer assumes that it is an elastic phenomenon and the equations required are conservation equations including Conservation of Mass (Continuity) and Conservation of Momentum or Conservation of Energy, and a relationship called the Equation of State which expresses the statistical stress condition of the body.

The distortion of the pipe, and liquid within, as it is subjected to a stress change is three-dimensional, but the analysis assuming an essentially one-dimensional (axisymmetric) behaviour provides an adequate basis for the solution of most problems. Suppose a sudden change (say a decrease) in the flow in the pipe, is produced. This means that, in a given period of time, more liquid is passing into a section than is leaving it and the container is strained to accommodate this. Thus, dynamically, the system sees an increased pressure force which is associated with the momentum change.

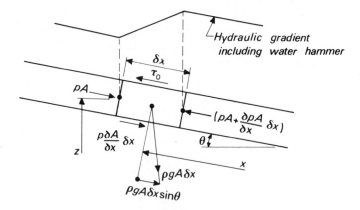

Fig. 2.1

In Fig. 2.1, treating momentum first, Newton's law provides

4

$$pA - \left(pA + \frac{\partial}{\partial x}(pA)\,\delta x\right) + p\frac{\partial A}{\partial x}\delta x + \rho g A \delta x \sin\theta - \tau_0 w_{\mathrm{p}}\delta x$$

$$= \rho A \delta x \frac{dv}{dt} \tag{2.1}$$

where p is the liquid pressure, increased by δp across an incremental length δx in a time δt. The average area of flow A is used for expressing the mass rate of flow of density ρ, which is moving at an average velocity v. The pipeline is inclined at an angle θ and τ_0 is the wall shear stress acting over a wetted perimeter w_{p}.

After dividing by the total mass $\rho A \delta x$, Equation (2.1) becomes

$$-\frac{1}{\rho}\frac{\partial p}{\partial x} + g\sin\theta - \frac{1}{\rho}\frac{\tau_0}{R} = \frac{dv}{dt} \tag{2.2}$$

where R is the ratio of the area divided by the wetted perimeter.

The pressure in the liquid may be written

$$p = \rho g(h - z)$$

where h is the pressure head at z above a datum. Assuming the change of ρ is small compared to the change in $(h - z)$, then

$$\frac{1}{\rho}\frac{\partial p}{\partial x} = g\left[\frac{\partial h}{\partial x} - \frac{\partial z}{\partial x}\right] \tag{2.3}$$

Substituting Equation (2.3) in (2.2) yields, since $\partial z/\partial x = -\sin\theta$,

$$g\frac{\partial h}{\partial x} + \frac{1}{\rho}\frac{\tau_0}{R} + \frac{dv}{dt} = 0 \tag{2.4}$$

Equation (2.4) in partial derivative form, ignoring the friction term, becomes

$$g\frac{\partial h}{\partial x} + v\frac{\partial v}{\partial x} + \frac{\partial v}{\partial t} = 0 \tag{2.5}$$

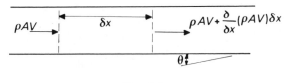

Fig. 2.2

In Fig. 2.2 the sudden change in mass flux associated with the strain of the container requires a continuity relation. Thus the change of mass with time in the control body δx in length must equal the net inflow, or outflow, of mass through the surface of the body. Hence, for unsteady flow:

$$\frac{\partial}{\partial t}(\rho A \delta x) = -\frac{\partial}{\partial x}(\rho A v)\delta x \tag{2.6}$$

Expanding and dividing by $\rho A \delta x$ yields

$$\frac{1}{\rho}\frac{\partial \rho}{\partial t} + \frac{1}{A}\frac{\partial A}{\partial t} + \frac{v}{\rho}\frac{\partial \rho}{\partial x} + \frac{\partial v}{\partial x} + \frac{v}{A}\frac{\partial A}{\partial x} = 0 \tag{2.7}$$

These may be grouped using the total derivatives:

$$\frac{dA}{dt} = \frac{\partial A}{\partial x}\frac{dx}{dt} + \frac{\partial A}{\partial t} \tag{2.8}$$

$$\frac{d\rho}{dt} = \frac{\partial \rho}{\partial x}\frac{dx}{dt} + \frac{\partial \rho}{\partial t} \tag{2.9}$$

Thus Equation (2.7) becomes

$$\frac{1}{A}\frac{dA}{dt} + \frac{1}{\rho}\frac{d\rho}{dt} + \frac{\partial v}{\partial x} = 0 \tag{2.10}$$

Equation (2.10) requires the use of a stress-strain relationship (an equation of state), to connect the deformation dA/dt and the stress change $d\rho/dt$. Both the container and the liquid itself undergo stress and strain changes. For the liquid

$$K = \rho\frac{dp}{d\rho} \tag{2.11}$$

where K is the bulk modulus of elasticity, whereas in the case of the solid the stress-strain conditions are more complex (see Appendix 1).

To show the overall physical behaviour, a simple expression for the

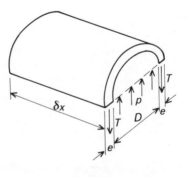

Fig. 2.3

container is derived, based on the hoop stress-strain realization in Fig. 2.3 for a circular pipe. With a pure hoop tension T,

$$\text{Strain} = \frac{p}{E}\frac{D}{2e} \tag{2.12}$$

where E is Young's elastic modulus, D is the diameter of the pipe and e is its wall thickness. The change in strain due to a change in δp produces a change in area for a circular pipe of $D^3\pi\,\delta p/4eE$. Thus, Equation (2.12) becomes

$$\frac{1}{A}\frac{dA}{dt} = \frac{D}{eE}\frac{dp}{dt} \tag{2.13}$$

Substitution of (2.11) and (2.13) in (2.10) gives

$$\frac{1}{\rho}\frac{dp}{dt} + C_0{}^2\frac{\partial v}{\partial x} = 0 \tag{2.14}$$

where the physical properties of the medium are connected by the constant

$$C_0{}^2 = \frac{K}{\rho}\left[\frac{1}{1 + \dfrac{D}{eE}K}\right] \tag{2.15}$$

Using $p = \rho g(h - z)$, Equation (2.14) may be expressed in terms of h, as follows:

$$\frac{dp}{dt} = \frac{\partial p}{\partial x}\frac{dx}{dt} + \frac{\partial p}{\partial t}$$

and

$$\frac{\partial p}{\partial x} = \rho g\left[\frac{\partial h}{\partial x} - \frac{\partial z}{\partial x}\right], \quad \frac{\partial p}{\partial t} = \rho g\left[\frac{\partial h}{\partial t} - \frac{\partial z}{\partial t}\right]$$

and thus

$$\frac{dp}{dt} = v\rho g\left[\frac{\partial h}{\partial x} - \frac{\partial z}{\partial x}\right] + \rho g\left[\frac{\partial h}{\partial t} - \frac{\partial z}{\partial t}\right] \tag{2.16}$$

Equation (2.14) then becomes

$$v\frac{\partial h}{\partial x} + \frac{\partial h}{\partial t} + \frac{C_0{}^2}{g}\frac{\partial v}{\partial x} + v\sin\theta = 0 \tag{2.17}$$

In arriving at (2.17) it was assumed the pipe was at rest and so $\partial z/\partial t = 0$.

2.2   Problems on the fundamental equations

2.2.1   *What do the fundamental equations mean in terms of the basic physics of the problem?*

The simultaneous pair of partial differential equations in their simplest form are

$$g\frac{\partial h}{\partial x} + \frac{\partial v}{\partial t} = 0 \tag{2.18}$$

$$\frac{C_0^2}{g}\frac{\partial v}{\partial x} + \frac{\partial h}{\partial t} = 0 \tag{2.19}$$

Differentiating each of the above equations, the following may be obtained:

$$\frac{\partial^2 v}{\partial t^2} = C_0^2 \frac{\partial^2 v}{\partial x^2} \tag{2.20}$$

and

$$\frac{\partial^2 h}{\partial t^2} = C_0^2 \frac{\partial^2 h}{\partial x^2} \tag{2.21}$$

The phenomenon of water hammer is thus confirmed as an elastic wave propagation.

2.2.2   *Discuss the limitations of the basic equations.*

The Equations (2.18) and (2.19) have omitted friction (see Equation (2.4)) and the terms $v\sin\theta$ and $v\partial v/\partial x$ and $v\partial h/\partial x$ have also been neglected. The term $v\sin\theta$ can be eliminated by taking the pipeline as horizontal but otherwise it is small if v is small compared to the wave speed.

Regarding the terms $v\partial v/\partial x$ and $v\partial h/\partial x$, if the wave motion consists of a discrete wave front which can be regarded as a discontinuity then, at a discontinuity which is moving at a velocity C_0,

$$x = \pm C_0 t + \text{constant} \tag{2.22}$$

The alternative sign recognizes that the general solution involves waves which may pass in either direction.

Since $dx/dt = \pm C_0$ at the wave front from Equation (2.22), then

$$v\frac{\partial v}{\partial x} = \pm \frac{v}{C_0}\frac{\partial v}{\partial t} \tag{2.23}$$

The value of v/C_0 will be quite small, no greater than 1/100 in most cases and so this term can be neglected in comparison to $\partial v/\partial t$. A similar argument applies to $v\partial h/\partial x$.

In the case of friction, the development of solution techniques for the graphical or numerical methods can proceed without its inclusion but it has profound implications when related to real valve operation and losses generally, as will be shown later. Steady state friction is treated comprehensively in most fluid mechanics texts for the case of axisymmetric flow. In the case of unsteady flow, however, there is a complex interaction in the fluid at and behind the surge front. In the case of gases, it has been observed (Holder, Stuart and North, 1961) that the losses and modification of the shock front are associated with boundary layer interaction. In water hammer in liquids the pressure wave does not usually consist of a thin interface, as in gases, but comprises a 'thick' transition region and the importance of boundary layer interaction and friction loss is not nearly as well understood. Friction is treated in special problems in due course.

2.2.3   What is the integral form of the basic equations?

The solution of partial differential equations involves arbitrary functions. The method of d'Alembert is used to integrate the equations and it may be shown (see Appendix 2), that the pressure head h and velocity v at time t are related as follows:

$$h - h_0 = F\left(t - \frac{x}{C_0}\right) + f\left(t + \frac{x}{C_0}\right) \tag{2.24}$$

$$v - v_0 = -\frac{g}{C_0}\left[F\left(t - \frac{x}{C_0}\right) - f\left(t + \frac{x}{C_0}\right)\right] \tag{2.25}$$

In these equations h_0 and v_0 represent an initial steady state condition and h and v conditions of unsteady flow. F and f are arbitrary functions with arguments in x and t. The physical significance of these functions may be interpreted as waves propagating in opposite directions and, at any point at any instant in time, conditions will be the sum of both.

2.2.4   Can the integral forms of the basic equations show the essential relationship between the h and v changes?

Extending the idea that the arbitrary functions can be regarded as wave propagations in opposite directions, Equations (2.24) and (2.25) may be arranged by addition and subtraction to become

$$h - h_0 = \frac{C_0}{g}(v - v_0) + 2F\left(t - \frac{x}{C_0}\right) \tag{2.26}$$

$$h - h_0 = \frac{-C_0}{g}(v - v_0) + 2f\left(t + \frac{x}{C_0}\right) \tag{2.27}$$

The F waves move in the negative velocity direction which corresponds to the positive x direction and the f waves move in the positive velocity direction which corresponds to the negative x direction. The symbolic association of waves with the arbitrary functions F and f is convenient since their direction in relation to velocity, once determined, is fixed and thus Equation (2.26) expresses events in terms of F waves and Equation (2.27) in terms of f waves. The analysis, however, requires the systematic elimination of the terms F and f in the above equations so the relationship between h and v may be known at any point at any time. This is achieved by determining certain fundamental properties of the F and f functions.

(1) Wave reflection Referring to Equation (2.24), at a reservoir, it is clear that only static head can exist, therefore:

$$h = h_0 + F\left(t - \frac{L}{C_0}\right) + f\left(t + \frac{L}{C_0}\right) = h_0$$

where L is the distance to the reservoir from a reference point. This establishes the very important property that an F wave is equal in magnitude but opposite in sign to an f wave at a reservoir.

$$F\left(t - \frac{L}{C_0}\right) = -f\left(t + \frac{L}{C_0}\right) \tag{2.28}$$

As these waves proceed in opposite directions, this is clearly the property of reflection at a reservoir.

(2) Wave constancy If an observer moves with the wave, at velocity C_0, then the equation for the observer is $x = x_1 + C_0 t$ where x_1 is his original position and he is moving in the positive x direction, i.e. an F wave. Hence

$$F\left(t - \frac{x}{C_0}\right) = F\left(t - \frac{x_1 + C_0 t}{C_0}\right) = F\left(\frac{-x_1}{C_0}\right) \tag{2.29}$$

Since this result is independent of x and t, the F wave appears constant to an observer who travels with the wave. The same is true for an f wave and thus the constancy of F and f waves is demonstrated.

With the establishment of these two properties, it is then readily shown that in general a wave must incorporate simultaneously pressure and velocity changes that are compatible with each other, and in magnitude correspond to

$$h - h_0 = \pm \frac{C_0}{g}(v - v_0) \tag{2.30}$$

A value of C_0 of the order of 1200 m s^{-1} is typical and thus a velocity change of 1 m s^{-1} produces 120 m of pressure head.

2.2.5 Can the energy principle be used to develop the basic equations?

The energy principle is given prominence in fluid mechanics texts for steady flow problems where it leads into the Bernoulli principle. The above treatment featured the momentum principle in developing the fundamental equations. The energy principle may be used (Wood and Stelson, 1966) and it is claimed that experimental pressure rises in excess of theoretical can be partly explained by a proper determination of a kinetic energy factor for the velocity distribution. It is of interest to note that the preceding momentum theory made no attempt to allow for a velocity distribution, although this can be done.

   The result of comparing the experimental work of Joukowski (1904) and Glover (1933) with the equation taking no account of the velocity distribution (Equation 2.30) shows experimental values greater by an order of 10% to 20%, being greater for lower initial pressure. However, a number of effects must be considered in addition to velocity distribution, i.e. the variation of the bulk modulus K for low pressures, head recovery because of pipe friction and three-dimensional aspects of the flow. The problem lies in defining the real fluid conditions during unsteady flow rather than the basic principle used in developing the equations and in this text the analysis is restricted to the simplified momentum theory.

2.3 Problems

2.3.1 What is the wave speed for the following cases?

See Appendix 1

Assume Poisson's ratio $(\mu) = 0.3$
(1) Water, rigid steel pipeline, diameter $(D) = 0.9$ m, wall thickness $(e) = 15$ mm   *Answer* 212 m s^{-1} $(T = 50°C)$
(2) Water, class C asbestos cement laid below ground, $D = 0.6$ m (nominal), $e = 46$ mm   *Answer* 1014 m s^{-1} $(T = 0°C)$
(3) Water, mild steel expansion joint pipeline laid above ground, $D = 0.8$ m, $e = 15$ mm   *Answer* 1160 m s^{-1} $(T = 0°C)$
(4) Oil, freely supported steel pipeline, $D = 0.1$ m, $e = 10$ mm (Assume $K = 1.5 \times 10^9$ N m^{-2} and $\rho = 900$ kg m^{-3})   *Answer* 1247 m s^{-1}

3

Graphical and numerical solutions

3.1 Graphical method

Water hammer problems can be solved graphically in the h, v domain. The simultaneous linear algebraic equations (2.26) and (2.27) are used with construction lines sloping $\pm C_0/g$. The graphical technique is sometimes referred to as the Schnyder and Bergeron graphical analysis (see Jaeger, 1956) although a number of significant contributions have been made by others, notably Angus (1935, 1938). Originally, the simultaneous equations (2.24) and (2.25) were referred to as the Allievi interlocked equations as a tribute to the great contribution made by L. Allievi (1913), mainly through exhaustive algebraic solutions of basic problems using these equations.

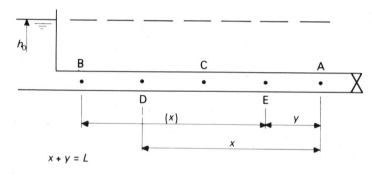

Fig. 3.1

In Fig. 3.1 a simple case of a pipeline bounded by a downstream valve and an upstream reservoir is illustrated. The convention of velocity positive towards the valve and distance x positive upstream is adopted and thus F and f waves always proceed upstream and downstream respectively. For a wave proceeding from A to B Equation (2.26) may be written as follows:

$$h_{A_{t_0}} - h_{A-0} = \frac{C_0}{g}\left(v_{A_{t_0}} - v_{A-0}\right) + 2F\left(t_0\right) \tag{3.1}$$

$$h_{D_{t_x}} - h_{D-0} = \frac{C_0}{g}\left(v_{D_{t_x}} - v_{D-0}\right) + 2F\left(t_x - \frac{x}{C_0}\right) \tag{3.2}$$

12

$$h_{B_{t_L}} - h_{B_{-0}} = \frac{C_0}{g}\left(v_{B_{t_L}} - v_{B_{-0}}\right) + 2F\left(t_L - \frac{L}{C_0}\right) \qquad (3.3)$$

In these equations, the following convention has been adopted:

Subscript A_{t_0} means time $t = 0$ at A

Subscript A_{-0} means time prior to $t = 0$ at A.

Since the waves are proceeding at a constant velocity C_0 the time for a wave to travel a distance x is x/C_0. Hence, the subscript D_{t_x} is a further abbreviation signifying time $t = x/C_0$ at D. Later it will be convenient for further abbreviation to use $L/C_0 = 1$ time unit. In all cases the time shown means at that instant a change has occurred (with the exception of subscript '-0') and the new condition will be sustained there until another designation appears (see later).

Equations (3.1) to (3.3) may be arranged by successive subtraction to yield:

$$h_{D_{t_x}} - h_{A_{t_0}} = \frac{C_0}{g}\left(v_{D_{t_x}} - v_{A_{t_0}}\right) \qquad (3.4)$$

$$h_{B_{t_L}} - h_{D_{t_x}} = \frac{C_0}{g}\left(v_{B_{t_L}} - v_{D_{t_x}}\right) \qquad (3.5)$$

and with further subtraction:

$$h_{B_{t_L}} - h_{A_{t_0}} = \frac{C_0}{g}\left(v_{B_{t_L}} - v_{A_{t_0}}\right) \qquad (3.6)$$

This operation has merely eliminated the F wave which is constant in proceeding from A to B and in a system where friction is ignored

$$h_{A_{-0}} = h_{D_{-0}} = h_{B_{-0}}$$

A similar procedure may be adopted using Equation (2.27) for a wave proceeding from B to A (note $AD = BE$) and consequently:

$$h_{E_{t_x}} - h_{B_{t_0}} = \frac{-C_0}{g}\left(v_{E_{t_x}} - v_{B_{t_0}}\right) \qquad (3.7)$$

$$h_{A_{t_L}} - h_{E_{t_x}} = \frac{-C_0}{g}\left(v_{A_{t_L}} - v_{E_{t_x}}\right) \qquad (3.8)$$

and thus

$$h_{A_{t_L}} - h_{B_{t_0}} = \frac{-C_0}{g}\left(v_{A_{t_L}} - v_{B_{t_0}}\right) \qquad (3.9)$$

Adapting this now to graphical analysis, any of the Equations (3.1) to

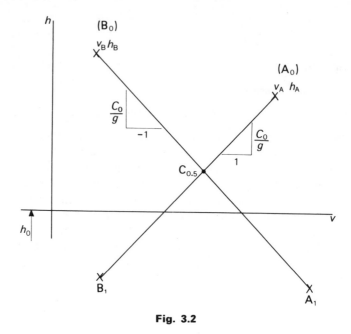

Fig. 3.2

(3.6) may be represented by straight lines with a slope of $+C_0/g$ and Equations (3.7) to (3.9) have a slope of $-C_0/g$, as shown in Fig. 3.2. Fig. 3.2 depicts waves simultaneously passing in opposite directions between A and B. Because the equation of wave motion is linear the principle of superposition holds and, consequently, the effect of these two waves may be added together to give a general solution. Thus if $x = y$ so that D and E both coincide with C then Equations (3.4) and (3.7) provide, respectively,

$$h_{C_t} - L_{A_{t_0}} = \frac{C_0}{g}\left(v_{C_t} - v_{A_{t_0}}\right) \tag{3.10}$$

and

$$h_{C_t} - L_{B_{t_0}} = \frac{-C_0}{g}\left(v_{C_t} - v_{B_{t_0}}\right) \tag{3.11}$$

where $t - t_0 = 0.5\,L/C_0 = 1/2$ units of time.

The implication of this result is very important. To find conditions at C at time t propagations from known conditions at points upstream and downstream are followed to arrive simultaneously at C. Proceeding upstream to C a slope $+C_0/g$ is required and downstream a slope of $-C_0/g$. It should be noted that the convention of velocity is positive downstream.

Initial conditions   All water hammer solutions are the result of an initial disturbance and the analysis will take account of everything that happens thereafter in its correct sequence in time, including any additional disturbances. The initiation of the waves may be related to some boundary condition which is applied such as a valve closure which introduces an unsteady flow in an otherwise steady state system. The accurate description of the various boundary conditions is the preoccupation of researchers on this subject and problems relating to them will now be treated.

3.1.1 Solve graphically an instantaneous total valve closure.

In Fig. 3.3a the valve at A is closed instantaneously. Friction is ignored and therefore in the h, v domain, as shown in Fig. 3.3b, all points from A to B are the designation A_{-0} for steady state conditions prior to any disturbance. It will be convenient to use $45°$ sloping lines for the characteristic F and f propagations and if the scale for velocity is 1 m s^{-1} per 10 cm then a scale of 100 m per 10 cm for h would mean $C_0/g = 100$ slopes at $45°$.

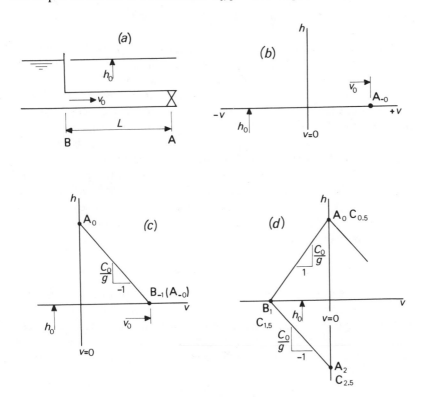

Fig. 3.3

The graphical analysis is the solution of the two simultaneous equations but initially only one equation is necessary if a boundary condition is specified. At B_{-1}, i.e. L/C_0 seconds in time before $t = 0$ at B, the boundary condition is known as an open reservoir with $h = h_0$ and $v = v_0$. Hence, B_{-1} is also at A_{-0}. Using Equation (3.9) with these new times:

$$h_{A_{t_0}} - h_{B_{t_{-1}}} = \frac{-C_0}{g} \left[v_{A_{t_0}} - v_{B_{t_{-1}}} \right] \tag{3.12}$$

This corresponds to the line in Fig. 3.3c at a slope $-C_0/g$ from B_{-1} to A_0, the latter point lying on the $v = 0$ line since at time $t = 0$ the gate valve was closed completely and initiated a wave, the magnitude of which is now known.

In this example the next event of interest is at B_1 when the first change arrives there and use of Equation (3.6):

$$h_{B_{t_1}} - h_{A_{t_0}} = \frac{C_0}{g} \left[v_{B_{t_1}} - v_{A_{t_0}} \right] \tag{3.13}$$

provides the line in Fig. 3.3d which locates B_1 on the $h = h_0$ line, since this is an open reservoir.

This example is very straightforward since only one wave is alternatively proceeding from A to B, then B to A. A_2 is readily located as shown in (d) and it should be clear that the designation A_0 now means at A from time $t = 0$ until just before $t = 2$. Provided every event has been recorded, then the conditions for any other part of the system can be recovered from the diagram. Thus, C (midpoint of AB) would experience pressure and velocity effects illustrated in (d) by the points $C_{0.5}$, $C_{1.5}$, and $C_{2.5}$. However, to locate these points both F and f propagations from known conditions to arrive at C simultaneously need to be followed. For example, $C_{1.5}$ must experience propagations which leave A_1 and B_1. It should be noted that A_1 is the same as A_0 since no new change occurs at A until A_2. The intersection of an F characteristic from $A_1 (A_0)$ which always has a slope of $-C_0/g$ is clearly at B_1 which is also, therefore, $C_{1.5}$.

3.1.2   Produce the time history of h and v for the valve closure.

In Fig. 3.4 the velocity time and pressure head time records for observers located at A and C are illustrated. The previous analysis is based on the fundamental equations of water hammer and requires a relation between pressure head and velocity according to Equation (2.30). In the problem shown in Fig. 3.3, it seems that the pressure head generated by a rapid change of v_0 to zero is obviously $\dfrac{C_0}{g} v_0$ and it would also seem that this would be readily observed in practice. It is important to realize that this is

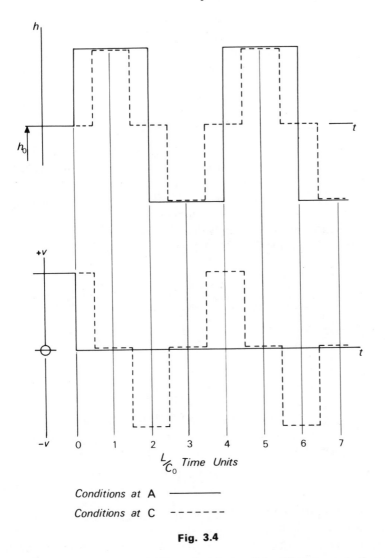

Conditions at A ————

Conditions at C - - - - - -

Fig. 3.4

the only time that such a confirmation is possible. The reason is that the events happen very quickly and as yet there has been limited success in measuring transient velocities of this kind (see Safwat, 1972). In Fig. 3.4, therefore, the velocity time diagrams are difficult to verify by experiment, whereas the pressure head time record can be obtained quite readily.

3.2 Numerical solutions

This text provides one application of the method of characteristics so that a useful set of computer programs can be developed to support a range of introductory questions and answers to water hammer problems. The basics of the method have been detailed in Appendix 3. The graphical solutions are given prominence in the main text because they readily relate to the physics of wave propagation.

3.2.1   *List a computer program for all points for Problem 3.1.1.*

The computer program in FORTRAN is based on the method of characteristics (see Appendix 3) (Fig. P.1). It will exclude the housekeeping of DIMENSION and COMMON statements, COMMENT cards, and DATA INPUT. It will be divided into SUBROUTINES for each task and will be applicable to NP pipes, although the solution to this specific problem is with $NP = 1$. Generally there will be $J = 1$ to NP pipes divided into $N(J)$ partitions giving $N(J) + 1$ grid points for each pipe.

```
        PROGRAM MAIN
C       **   DIMENSION COMMON STATEMENTS, READ OF DATA  **
C       **   MM IS NUMBER OF DT INCREMENTS, PL=PIPE LENGTH (L)   **
C       **   Q=DISCHARGE, D=PIPE DIAMETER, CO=WAVE SPEED  **
        CALL INIT(.........),
        DO 100 JK=1,MM
        CALL TERM(........)
        CALL INTER(.......)
        DO 99 J=1,NP
        NOP=N(J)+1
        DO 99 I=1,NOP
        V(J,I)=VP(J,I)
        H(J,I)=HP(J,I)
99      CONTINUE
C       **   OUTPUT RESULTS V,H FOR SELECTED J,I   **
100     CONTINUE
        CALL EXIT
        END
```

Fig. P.1

SUBROUTINE INIT (Fig. P.2) provides for the determination of the interpolation ratios R(J), the time increment DT and initial H and V. SUBROUTINE TERM (Fig. P.3) caters for both upstream and downstream reservoirs whereas SUBROUTINE INTER (Fig. P.4) calculates all interior points. In this case with one pipe ON would be equal to the number of partitions, chosen to give detail that might be appropriate to a pipeline over changing terrain, whereas ON would normally be put equal to 2 for the shortest pipe in a more complex system. HSTAT is the level of the downstream reservoir.

There will be a program MAIN (Fig. P.1) which will do the housekeeping referred to above and which will set up the iteration loops required to determine the development of the water hammer with time. The time

```
        SUBROUTINE INIT(........)
        DO 30 J=1,NP
        AN(J)=PL(J)/CO(J)
        IF(AN(1).LE.AN(J)) GO TO 30
        AN(1)=AN(J)
30      CONTINUE
        DT=AN(1)/ON
        DO 40 J=1,NP
        AR(J)=0.78539816*D(J)**2
        FK(J)=F(J)*DT/(2.*D(J))
        HF(J)=F(J)*PL(J)*Q(J)*ABS(Q(J))/(2.*G*AR(J)**2*D(J))
        AN(J) = PL(J)/(DT*CO(J))
        R(J) =AN(J)*DT/PL(J)
        N(J)=AN(J)
40      CONTINUE
        HFE=HSTAT
        DO 42 J=1,NP
        NOP=N(J)+1
        HFI=HF(J)/N(J)
        H(J,1)=HFE
        V(J,1)=Q(J)/AR(J)
        DO 42 I=2,NOP
        H(J,I)=HFE+HFI
        V(J,I)=Q(J)/AR(J)
        HFE=HFE+HFI
42      CONTINUE
        RETURN
        END
```

Fig. P.2

interval will be DT which is the shortest pipe length divided by the wave speed and then divided by ON, provided there is no other smaller ratio of L/C_0. The program MAIN will also handle the output which might be a tabulation of values or a scheduling for a graphical OUTPUT. The program of course calls INIT once and then each of the SUBROUTINES it requires with each iteration.

In the SUBROUTINE INIT the old values H and V are first established.

```
        SUBROUTINE TERM(.........)
        J=1
        I=1
        VS = V(J,I)-R(J)*CO(J)*(V(J,I)-V(J,I+1))
        HS = H(J,I)-R(J)*CO(J)*(H(J,I)-H(J,I+1))
        VSS=VS+G/CO(J)*HS-FK(J)*VS*ABS(VS)
        VP(J,I)=0.
        HP(J,I)=(VSS-VP(J,I))*CO(J)/G
        J=NP
        I=N(J)+1
        VR = V(J,I)-R(J)*CO(J)*(V(J,I)-V(J,I-1))
        HR = H(J,I)-R(J)*CO(J)*(H(J,I)-H(J,I-1))
        HP(J,I)=H(J,I)
        VP(J,I)=VR+G/CO(J)*(HP(J,I)-HR)-FK(J)*VR*ABS(VR)
10      CONTINUE
        RETURN
        END
```

Fig. P.3

```
SUBROUTINE INTER(.........)
DO 10 J=1,NP
NOP=N(J)
DO 10 I=2,NOP
IF(PL(J).EQ.0.) GO TO 10
VR = V(J,I)-R(J)*CO(J)*(V(J,I)-V(J,I-1))
HR = H(J,I)-R(J)*CO(J)*(H(J,I)-H(J,I-1))
VS = V(J,I)-R(J)*CO(J)*(V(J,I)-V(J,I+1))
HS = H(J,I)-R(J)*CO(J)*(H(J,I)-H(J,I+1))
VP(J,I)=(VR+VS-G/CO(J)*(HR-HS)-FK(J)*(VR*ABS(VR)+VS*ABS(VS)))/2.
HP(J,I)=(HR+HS-CO(J)/G*(VR-VS+FK(J)*(VS*ABS(VS)-VR*ABS(VR))))/2.
10    CONTINUE
RETURN
END
```

Fig. P.4

Calculations then determine for an increment DT the new or present values HP and VP for every grid point. These now become the old values prior to the next time increment and hence a small DO LOOP is the last operation required to replace all these values as H and V.

FRICTION is incorporated in the above solution but a comparison can be made with the no-friction graphical solution by putting all $F(J) = 0$. In this problem the instantaneous valve closure is simulated by setting $VP(1,1) = 0$ in SUBROUTINE TERM.

4

Elementary gate characteristic (linear gate)

4.1 Introduction

The gate valve is treated first in the absence of pipe friction for historical reasons, although many misconceptions can arise from this oversimplification. The boundary conditions assigned to this idealized device should be regarded as a means of convenience for producing a (reader) controlled velocity change for the purpose of generating wave propagations, so that greater understanding of wave mechanics can be achieved. The realistic non-linear gate is considered later, after problems related to friction are discussed.

The static pressure head available to cause a velocity v_0 in a gravity pipeline can be wasted conveniently in a nozzle at the downstream end. The gate operation can be regarded as a variation of the nozzle from a fully open to a closed position in a series of finite steps or time intervals which may be as small as one cares to choose.

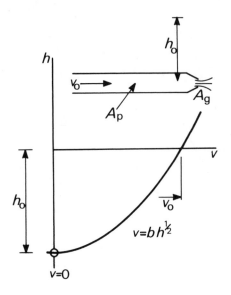

Fig. 4.1

21

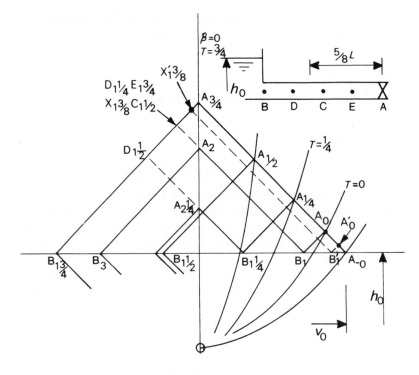

Fig. 4.3

X will experience the maximum water hammer whereas points upstream will experience less, in a non-linear manner, until there is zero water hammer pressure head at B.

4.2.3   Solve for a gate closure in greater than $2L/C_0$.

Fig. 4.4 shows a general case of gate closure in 4 steps in a time of $3L/C_0$. The number of steps could be as low as 2 for this solution ($\beta = \frac{1}{2}$ and $\beta = 0$) and all necessary information would be incorporated, but clearly additional steps smooth out the solution and would in fact possibly indicate larger water hammer valves during the closure process. Thus the head at A_2 is slightly greater than at A_1 or A_3. It should be noted that, unless otherwise specified, $L/C_0 = 1$ time unit.

4.2.4   Solve for the case of a downstream gate opening in $3L/C_0 s$ in 4 steps.

Fig. 4.5 shows the solution for the gate opening case. The initial conditions everywhere are at A_{-0} at zero velocity and head h_0. At A_0 a new gate

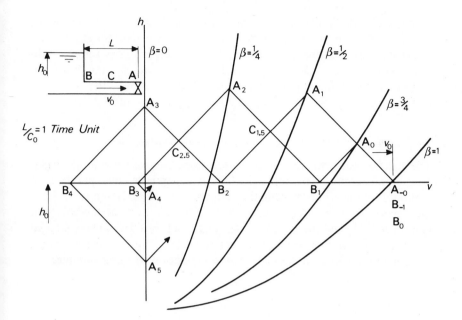

Fig. 4.4

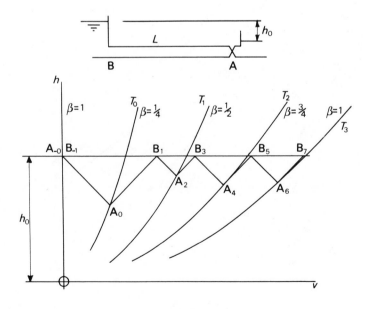

Fig. 4.5

position is set and proceeding from known conditions at B_{-1} in the positive velocity direction, an f type slope of $-C_0/g$ provides the appropriate intersection on the gate characteristic. The solution proceeds automatically as in Problem 4.2.3. Each gate characteristic is a boundary condition one chooses and thus any type of non-uniform gate position with time can be selected fulfilling the requirement that it is fully open in $3L/C_0$ s.

4.2.5   Outline a computer solution for gate operation.

The general treatment is left until the realistic non-linear behaviour of a gate is developed. It should be emphasized that in general the valve characteristic without friction as illustrated in this chapter is of academic interest as far as the valve operation itself is concerned. The gate as treated above was conceived by Allievi and it is useful, as mentioned on p. 21, but current texts that provide computer solutions close to the Allievi linear solution do little to explain the real complexity of valves in relation to water hammer.

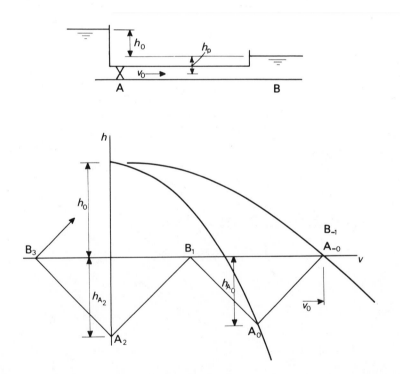

Fig. 4.6

4.2.6   Describe conditions for a gate closure when it is situated at the upstream end of a pipeline.

In Fig. 4.6 the gate is now upstream and v is positive downstream from A so that an observer follows an f type propagation in moving from A to B or an F type propagation from B to A. The gate characteristic for the no friction case drops so that A_{-0} appears h_0 below the upstream reservoir since it must be assumed here that h_0 is dropped across the gate in the steady state condition. A_0 on the arbitrary two step closure characteristic is found by an observer leaving B_{-1} along an F propagation which has a $+C_0/g$ slope and thence an f propagation returns to B_1.

The importance of this solution is the fact that the pressures may become less than atmospheric in the pipeline if h_{A_0} or h_{A_2} exceed h_p in magnitude. This not only suggests design problems for large diameter pipelines it also introduces the possibility of water column rupture, i.e. separation of the water column because it cannot sustain tension. This result indicates the general unsuitability of an upstream control in a system and even a gate for isolation purposes must be operated very carefully and if possible when little flow is taking place.

5

The pump characteristic

5.1 Introduction

As a boundary condition responsible for exerting a control in connection with the propagation of dynamic pressures, a pumping plant is of major importance. It is largely because of the manufacture of high speed small inertia prime movers and the parallel advance in the design of more compact pumps that the analytical interest in water hammer and surge problems is no longer almost exclusively restricted to hydroelectric practice. As pumps and their prime movers no longer take minutes to slow down when stopped either deliberately or accidentally, the resultant pressures for pumping mains become an important design problem. In the case of electrically powered plant the possibility of a power failure means that there is no safeguard at all against a pump suddenly failing to discharge. Unlike a hydroelectric scheme, the pumping system does not require an exacting control of the speed of the machines and therefore some fluctuations are permissible, more than would be tolerated in the former. It is customary to allow for conditions of rapid stopping or starting of pumps.

The H–Q diagram for a pump may be readily converted to h–v for any chosen system using the area of the main pipeline for calculating the velocity. The subsidiary pipework adjacent to the pump, generally of a smaller diameter, is usually so short as to be ignored in relation to the pumping main. In Fig. 5.1 the pump characteristic for a constant speed is

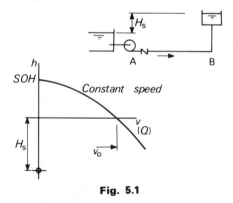

Fig. 5.1

28

shown and, if a non-return (reflux) valve is installed on the delivery side, the line at level H_s represents the hydraulic gradient for the pipeline AB when friction is ignored. SOH is the shut-off-head of the pump which in this case is a centrifugal or mixed flow type. The line through zero head represents the level of the storage on the suction side. The scales of the diagram for the pump characteristic may be easily chosen so that the C_0/g lines slope at $45°$ for convenience in performing a graphical analysis.

The point A at the upstream or pump end may represent the pump when there is flow and the non-return valve is open, or it may be considered just to the right of the non-return valve when it is closed. In the latter case this is because it is the pipeline AB that is being analysed. If the distance from the suction reservoir to A is much less than AB, then the reservoir is also considered as A when the non-return valve is open and if the pump is considered as not impeding the flow. The non-return valve is assumed to have the property of no resistance to flow when it is in the positive AB direction and it does not allow negative (reverse) flow to occur at A.

Initially the pump characteristic will be considered as a boundary condition which can be chosen as desired to illustrate the effect on wave propagations. Machine inertia will be considered later.

5.2 Problems

5.2.1 Show graphically the condition for a pump start – the rated speed being achieved instantaneously.

In Fig. 5.2 with velocity positive from A to B, known conditions at B_{-1} are propagated with an F type propagation at slope $+C_0/g$ toward A. Since the non-return valve prevents reverse flow initial conditions everywhere downstream of A are at H_s and $v = 0$ and thus from B_{-1}, A_0 and subsequent conditions are readily located as shown. Friction has been ignored and the maximum water hammer will in general not exceed the shut-off-head of the pump.

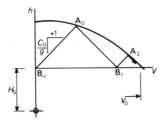

Fig. 5.2

5.2.2 Assume the pump characteristic is of secondary importance and analyse the pump failure case.

The stopping of a pump results in a number of alternative conditions depending on the pump head and associated control valves. The failure of a pump to maintain supply normally involves the operation of a non-return valve or some delayed valve closing to prevent the draining of the pumping main. In most cases, the subsequent behaviour of the pump, which might operate as a turbine with reverse flow through it, is an important control feature and, as considered later, its inertia will also be a deciding factor.

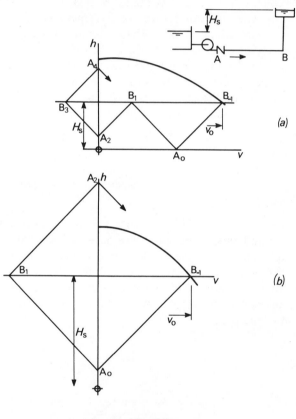

Fig. 5.3

Ignoring any effect that the pump inertia might have, Fig. 5.3 shows two examples, the result being determined by the amount of pump head. In (a) the boundary condition at A_0 met by an observer leaving B_{-1} will be the suction tank level. At A_2 the non-return valve prevents reverse flow and then the solution cycles as shown, the non-return valve remaining closed. In (b) the observer meets the boundary condition of no reverse flow before the pressure can fall as low as the suction tank. In both these cases it is essential that the non-return valve c perates. The correct operation of the non-return valve is required now and at any future time if these are the design conditions that will be provided for in specifying the pressures in the pipeline.

If there is a malfunction of the non-return valve which leads to a delayed closing, conditions might be as illustrated in Fig. 5.4. When flow is in reverse through the machine the boundary condition at A can be represented by a nozzle loss, that is just like a gate characteristic (with no friction). Subsequent results will then depend on if and when the non-return valve operates. If it continues to stay open then the result is A_2' and convergence upon A_T. If after time 2 the non-return valve closes then the pressures corresponding to A_2'' would occur and this could be calamitous unless provision has been made for it in design.

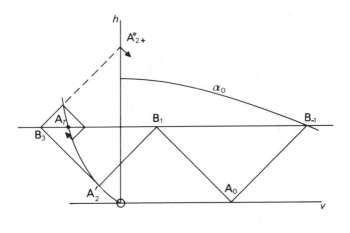

Fig. 5.4

5.2.3   Show the effect of a pump decelerating in times greater than $2L/C_0$.

In the previous cases for the failure of a pump to maintain discharge the associated control valve behaviour was highlighted. When a pump decelerates in times greater than $2L/C_0$, then the pump characteristic at different speeds (N) is required as shown in Fig. 5.5. In this case it is clear that the fall in pressure is not as great and hence, with proper control valve function, the subsequent overpressures will not be as severe.

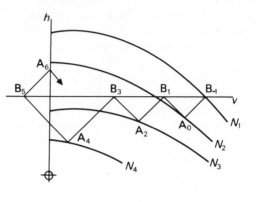

Fig. 5.5

5.2.4   *Examine the conditions of pump failure allowing for inertia.*

In Fig. 5.6, some typical data representing steady state conditions for a medium size centrifugal pump are included. The curves show the variation of h (head), q (discharge) and T (torque) for a number of speeds N, made dimensionless with respect to the rated Q_0, H_0 corresponding to the best efficiency point η_0 at a speed of N_0. The connecting relation for dimensionless torque is:

$$\beta = \frac{Q_R H_R}{\alpha E} \tag{5.1}$$

in which $Q_R = q/Q_0$, $H_R = h/H_0$, $\alpha = N/N_0$ and $E = \eta/\eta_0$.

It is useful to recall that points may be related to each other along iso-efficiency lines by the proportionalities:

$$Q_R \propto \alpha, H_R \propto \alpha^2 \quad \text{and} \quad \beta \propto \alpha^2 \tag{5.2}$$

It is generally assumed that transient conditions, i.e. whilst a pump is rapidly changing speed, can be extracted from the data shown in Fig. 5.6, provided the speed at a particular instant is known.

The change in speed of a machine may be found by using the basic equation for torque as follows:

$$T = -I\frac{d\omega}{dt} \tag{5.3}$$

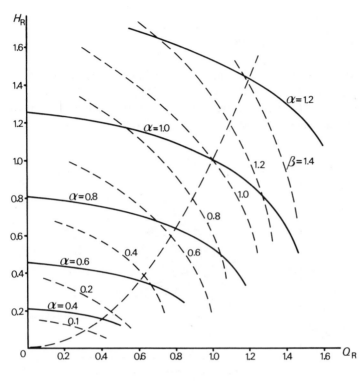

Fig. 5.6

where I is the second moment of the area of the rotating part and $d\omega/dt$ is the angular acceleration. In a finite small time interval Δt, the average torques T_1 and T_2 corresponding to ω_1 and ω_2 require:

$$\frac{T_1 + T_2}{2} = -\frac{GD^2}{4g}\left(\frac{\omega_2 - \omega_1}{\Delta t}\right) \qquad (5.4)$$

where I has been expressed in terms of GD^2, the flywheel effect of the rotating machine parts and entrained water. ($I = WR^2/g$ is customary in Imperial units).

Equation (5.4) may be rewritten to use the dimensionless ratios as follows:

$$\alpha_1 - \alpha_2 = M(\beta_1 + \beta_2)\Delta t \qquad (5.5)$$

where

$$M = \frac{1790\,000\,H_0 Q_0}{GD^2 \eta_0 N_0^2} \quad \text{(kg, m)} \qquad (5.6a)$$

and

$$M = \frac{91\,600\,H_0\,Q_0}{WR^2\,\eta_0\,N_0{}^2} \quad \text{(lbs, ft)} \tag{5.6b}$$

The procedure involves selecting a value for the change in torque, so that the associated speed change from Equation (5.5) provides at the new speed a value of v, h (or Q, H) corresponding to the transient change propagated by a water hammer characteristic at the appropriate time.

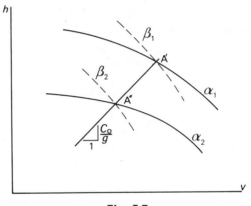

Fig. 5.7

Thus in Fig. 5.7, proceeding with a graphical solution, Equation (5.5) must be satisfied simultaneously with the requirement that:

$$h_{A'} - h_{A''} = \frac{C_0}{g}(v_{A'} - v_{A''}) \tag{5.7}$$

where A'' is Δt seconds later than A'.

The simultaneous solution of Equations (5.5) and (5.7) may be by an iteration procedure, whereas others (Linton, 1954; Streeter, 1963) have developed explicit solutions.

Very few tests have been conducted on machines to determine their total characteristic data and it is customary to choose the closest suitable approximation from the representations for three divergent specific speed types found by Knapp (1937). An estimate for the value of GD^2 is also often used since it must incorporate the combined flywheel effects of the pump (or turbine), the motor (or generator), the couplings and the enclosed liquid.

In practice, the GD^2 of an installed machine may be found by measuring its stopping time from a known speed. Of course, it is necessary to isolate the machine hydraulically from the pipeline for such a test. An approximate formula corresponding to American design pumps (Linton, 1954) is:

$$GD^2 = 912 \left(\frac{P_0}{N_0}\right)^{1.435} \tag{5.8}$$

where P_0 is the rated power in kW and N_0 is in rpm. This equation has been derived from data applying to a range of power from 120 to 1500 kW and speeds varying from 450 to 1800 rpm.

Ignoring friction, consider the case of a pipeline which is 0.5 m in diameter and 300 m long with $C_0 = 1200 \text{ m s}^{-1}$. The pump data are as follows:

$Q_0 = 0.3 \text{ m}^3 \text{ s}^{-1}$
$H_0 = 36 \text{ m}$
$\eta_0 = 0.85$
$N_0 = 1500 \text{ rpm}$

Calculations yield a power (P_0) of 106 kW and thus $GD^2 = 20.4$ from Equation (5.8) and $M = 0.505$ from Equation (5.6).

In Fig. 5.8 the data from Fig. 5.6 has been converted to an h, v representation with scales which require the characteristic slopes to be laid at $\pm 68°$. $2L/C_0$ is 0.5 s and a time not greater than this can be chosen for Δt and a value of 0.25 will be used.

Using trial and error to find α_2 in Equation (5.5) with $\alpha_1 = \beta_1 = 1$ and trying $\beta_2 = 0.6$ requires $\alpha_2 = 0.80$ which is point X_1 whereas $\beta_2 = 0.70$ gives $\alpha_2 = 0.787$ and X_2, neither of which lies on the characteristic through B_{-1} on a slope of $+C_0/g$. A further trial of $\beta_2 = 0.68$ gives $\alpha_2 = 0.795$ and the required condition A_0. According to the convention adopted A_0 is the instant a change occurs at A and it is the intersection of the pump characteristic produced by a change of speed corresponding to a time interval of 0.25 s and the propagation along a $+C_0/g$ characteristic (F type, v positive from A to B).

A further propagation must be considered leaving B to arrive at the pump at time L/C_0 at A, i.e. A_1. This is found in the same manner by a trial of $\beta_2 = 0.52$, so $\alpha_2 = 0.795 - 1.20/8 = 0.645$. The propagation from A_0 to B_1 must intersect the boundary of an unchanged static head at B and returns to A_2, $2L/C_0$ (0.5 s) later.

A further trial and error procedure is used to find A_2 with a further change in speed occurring, resulting in $\beta_2 = 0.35$ and $\alpha_2 = 0.53$. A_3 is also shown with $\beta = 0.25$, $\alpha = 0.46$. It is evident that the fall in pressure head in the system will be a maximum at A_6. Subsequent pressures, with proper operation of the reflux valve, oscillate about $v = 0$ and $h = H_0$. The minimum level of 30 m below the delivery tank level compares with 36 m at P which would have occurred if the pump inertia had been ignored.

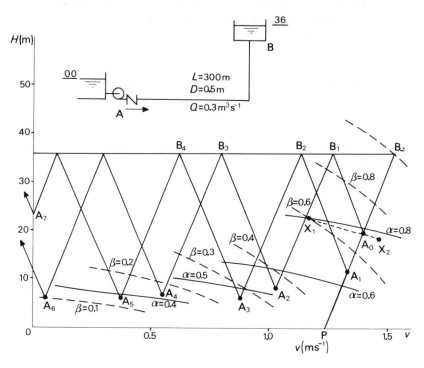

Fig. 5.8

5.2.5   Develop a computer SUBROUTINE for the normal zone of pump operation.

Appendix 3 describes the finite difference equations for analysis by the method of characteristics. The SUBROUTINE PUMP (Fig. P.5) will calculate the new values of H and V at the pump flange after the predetermined time increment DT. The associated pipeline is on the downstream side and so the pump requires a characteristic C^- to propagate conditions from the nearby interior point to the pump. The pump is at the upstream end of NP series of pipes and when the velocity goes to zero, a non-return valve acts to prevent reverse flow. The pump data is provided for equally spaced values of V/α and any value can be found by interpolation. Stirling's formula was used for values of H whereas a Gauss cubic interpolation formula was used for deriving values of torque.

```
        SUBROUTINE PUMP(..........)
C       ** HN=NORMAL PUMP HEAD, VV=INITIAL VELOCITY  (NORMAL)  **
C       ** AI AND VI PROVIDE PREVIOUS THREE VALUES OF SPEED AND VELOCITY
        J=NP
        I=N(J)+1
        VR=V(J,I)-R(J)*CO(J)*(V(J,I)-V(J,I-1))
        HR=H(J,I)-R(J)*CO(J)*(H(J,I)-H(J,I-1))
        VRR=VR-G/CO(J)*HR-FK(J)*VR*ABS(VR)
        AA=3.*AI(1)-3.*AI(2)+AI(3)
        VT=3.*VI(1)-3.*VI(2)+VI(3)
44      CONTINUE
        VX=VT/VV
        IF(AA.LE.0.) GO TO 80
        IF(VX) 49,49,50
49      GO TO 80
50      CONTINUE
        RX=VX/AA+1.
        L=ABS(RX)
        RR=RX-L
        IF(L.LT.2) GO TO 80
        H1=HAN(L-1)
        H2=HAN(L)
        H3=HAN(L+1)
        B1=BAN(L-1)
        B2=BAN(L)
        B3=BAN(L+1)
        B4=BAN(L+2)
51      CONTINUE
        BX=B3-B2
        BY=B4-B3-B2+B1
        BZ=B4-3.*B3+3.*B2-B1
        BB=(B2+RR*BX+RR*(RR-1.)*BY/4.+RR*(RR-.5)*(RR-1.)*BZ/6.)*AA**2
        AB=AP-P*DT*(BP+BB)
        X=H1*AB**2*HN
        Y=H2*AB**2*HN
        Z=H3*AB**2*HN
        AXX=HN*AB/10.
        VIP=VRR*CO(J)/G/AXX+1.-L
        A=(Z-2.*Y+X)/2./AXX**2
        B=(Z-X)/2./AXX+(Z-2.*Y+X)/AXX*VIP-1.
        C=Y+(Z-X)/2.*VIP+(Z-2.*Y+X)/2*VIP**2
        SQ=B**2-4.*A*C
        IF(SQ.LT.0.)GO TO 80
        IF(A.EQ.0)GO TO 53
        HX=(-B-SQRT(SQ))/2./A
        GO TO 90
53      CONTINUE
        HX=-C/B
        GO TO 90
80      CONTINUE
        AP=0.
        VP(J,I)=0.
        HP(J,I)=-VRR*CO(J)/G
        GO TO 93
90      CONTINUE
        AP=AB
        BP=BB
        HP(J,I)=HX
        VP(J,I)=HP(J,I)*G/CO(J)+VRR
93      CONTINUE
        RETURN
        END
```

Fig. P.5

5.2.6   Solve by computer Problem 5.2.4 and show the effect of varying the value of GD² by a factor of 2.

In Fig. 5.9 the results are shown graphically for the computer solutions of Problem 5.2.4 and for the two cases of twice and half the values of GD^2. The minimum and maximum values of pressure head are shown as well as the elapsed times in L/C_0 time units, where T_2 is for double the GD^2 and $T_{1/2}$ for the lesser value. The results clearly show the usefulness of sensitivity analyses for variables such as GD^2 which may only have an approximate value.

5.2.7   Is there an optimum location for a pumping plant with both suction and delivery mains?

The water hammer associated with propagation both upstream and downstream from a control point involves important principles which will

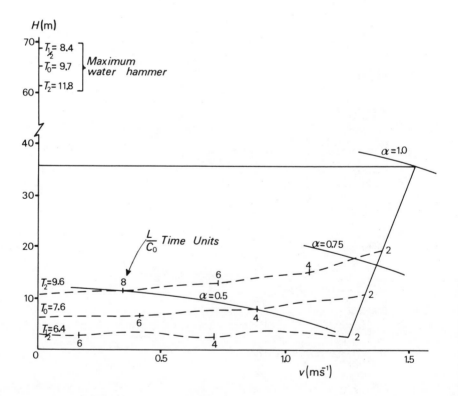

Fig. 5.9

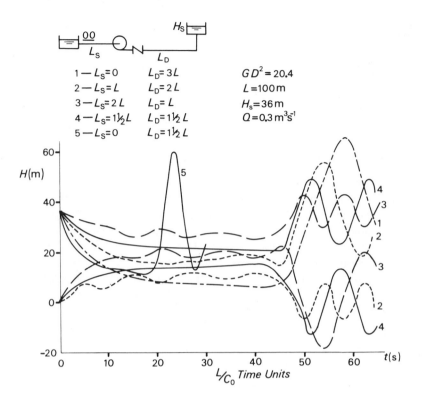

Fig. 5.10

be discussed later. For now, this question can be solved for a pumping installation using a computer program designed for the purpose. The comparisons will show not only the effect of varying the length of the suction main, but also the error in ignoring it completely. The cases analysed are shown in Fig. 5.10, where finite lengths are allowed on each side of the pump except the one where the suction is ignored completely, even though it equals the length on the delivery side.

The water hammer on the delivery side becomes less as it is shortened, but this is compensated by an increase on the suction side. When the pipelines are more nearly equal on both sides, the optimum would appear to be reached. It is also often found that the pump takes longer to decelerate on power failure with this arrangement and this would tend to reduce the water hammer also. If the suction side was ignored completely the water hammer is clearly overestimated as shown. These results suggest that in the planning stage the plant location might well be chosen with advantage to reduce the dynamic loading for the pipeline design.

6

Propagation simultaneously upstream and downstream

6.1 Introduction

Water hammer is not a one way transmission. Just as it is inconceivable that pressure in a fluid at a point acts in only one direction, so a disturbance in a pipe must be propagated in all directions equally. Hence in a pipe in the longitudinal (one-dimensional) sense considered here, a disturbance would propagate equally upstream and downstream. The physical bases are best established in general terms by considering that pipes upstream and downstream of a valve, for example, are of different areas and lengths. In using the graphical analysis it is rendered compact by normalizing all velocities to one reference value.

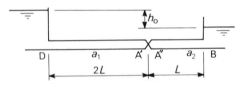

Fig. 6.1

In Fig. 6.1, ignoring friction, we must assume that h_0 is dropped across the fully open gate A. The conditions just upstream of the gate will be represented by A′ and downstream by A″. The pipe lengths are $2L$ and L with areas a_1 and a_2 respectively for AD and AB. Continuity yields for point A:

$$v_{A'} a_1 = v_{A''} a_2 \qquad (6.1)$$

For convenience in simplifying the construction, the h–v plot will use the velocity in AD and thus from Equation (6.1) the velocity in AB is scaled so that:

$$V_{A'} = v_{A'} \quad \text{and} \quad V_{A''} = v_{A''} \frac{a_2}{a_1}$$

A change of ΔV in AD produces a pressure head of $\dfrac{C_0}{g}\Delta V$. In AB, however,

$\dfrac{C_0}{g} \dfrac{a_1}{a_2} \Delta V$ is required for a change ΔV because of the scaling of velocities. The correct solution is readily obtained by using a slope for the propagation characteristic of $\dfrac{C_0}{g} \dfrac{a_1}{a_2}$ for AB. It should be noted that the h–v plot may be made dimensionless by using velocity ratios and slopes for the propagation characteristics of:

$$\pm \frac{C_0}{2g} \frac{v}{h}$$

but in this text, this is not used as it leads to difficulties with low pressures.

A rigorous solution for this case can now be developed as follows:
In AB:

$$
\begin{aligned}
h_{A''} &= \frac{C_0}{g}\left[v_{A''} - v_{A''_{-0}} \right] + 2\Sigma F \\
&= \frac{C_0 a_1}{g\, a_2}\left[V_{A''} - V_{A''_{-0}} \right] + 2\Sigma F
\end{aligned}
\tag{6.2}
$$

In AD:

$$
h_{A'} - h_{-0} = \frac{-C_0}{g}\left[V_{A'} - V_{A'_{-0}} \right] + 2\Sigma f
\tag{6.3}
$$

Hence, using $V_{A'} = V_{A''}$ and $V_{A'_{-0}} = V_{A''_{-0}}$

$$
h_{A'} - h_{A''} = h_{-0}\frac{-C_0}{g}\left[1 + \frac{a_1}{a_2} \right]\left[V_{A'} - V_{A''_{-0}} \right] + 2\Sigma f - 2\Sigma F
\tag{6.4}
$$

It should be noted that, for conditions at A, only type f in AD and type F in AB need to be considered.

In Fig. 6.2 the first effect for a gradual gate closure is represented by the gate characteristic, which is arbitrarily located for this example, at T_0. The analysis is drawn for A' as far as the representation of the gate is concerned. To find A_0 propagations from B_{-1} and D_{-2} are followed along characteristic slopes $\dfrac{C_0 a_1}{g\, a_2}$ and $\dfrac{-C_0}{g}$ according to Equations (6.2) and (6.3) respectively. Using these equations at this time eliminates the terms $2\Sigma f$ and $2\Sigma F$ since at B_{-1} and D_{-2} steady state conditions prevailed. Thus Equation (6.4) becomes:

$$
h_{A'_0} - h_{A''_0} = h_{-0} - \frac{C_0}{g}\left[1 + \frac{a_1}{a_2} \right]\left[V_{A'_0} - V_{A''_{-0}} \right]
\tag{6.5}
$$

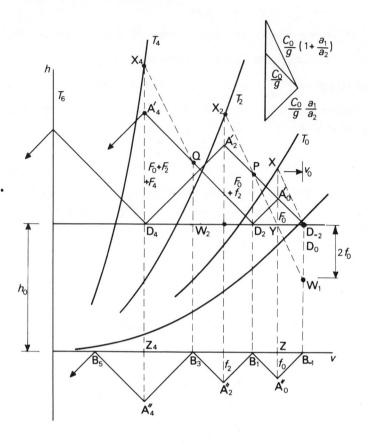

Fig. 6.2

The solution is found in Fig. 6.2 by using a characteristic of slope

$$\frac{-C_0}{g}\left[1+\frac{a_1}{a_2}\right]$$

to intersect the gate characteristic at X. It is readily seen that XY is the sum of YA'_0 and ZA''_0 so that $h_{-0} + XY = h_{A'_0} - h_{A''_0}$ as required. From A'_0 and A''_0, B_1 and D_2 are found by following the characteristic slopes $\dfrac{-C_0}{g}\dfrac{a_1}{a_2}$ and $\dfrac{C_0}{g}$ respectively and satisfying the fixed boundary conditions at B and D corresponding to the reservoir levels. Proceeding now to the next time interval (which should be not greater than $2L/C_0$) this is shown by a new gate position T_2. Propagations from B_1 and D_0 are required along characteristic slopes $\dfrac{C_0}{g}\dfrac{a_1}{a_2}$ and $\dfrac{-C_0}{g}$. Equation (6.4) may be used again but this time, and at later times, Σf and ΣF will not in general be zero.

In Fig. 6.3 the progression of waves in the system, symbolized by F_0, F_2, F_4, f_0, f_2 and f_4, is illustrated where $\Sigma(F+f)$ is the head variation and $\Sigma(F-f)$ is proportional to the velocity variation. The evaluation of $2\Sigma f$ and $2\Sigma F$ is shown and it is found at time $2L/C_0$ that:

$$h_{A_2'} - h_{A_2''} = h_{-0} - \frac{C_0}{g}\left[V_{A_2'} - V_{A''_{-0}}\right] - \frac{C_0}{g}\frac{a_1}{a_2}\left[V_{A_2'} - V_{A''_{-0}}\right] - 2f_0$$

$$= h_{-0} - \frac{C_0}{g}\left[1 + \frac{a_1}{a_2}\right]\left[V_{A_2'} - V_{A''_{-0}}\right] - 2f_0 \qquad (6.6)$$

Inspection shows that f_0 is symbolically the ordinate ZA_0'' and thus the construction to find A_2 requires a characteristic of slope $\dfrac{-C_0}{g}\left[1 + \dfrac{a_1}{a_2}\right]$ through the point W_1 in Fig. 6.2, intersecting the gate characteristic T_2 at X_2, whence the points A_2' and A_2'' are found by dropping the vertical to intersect the characteristic from D_0 and B_1 respectively.

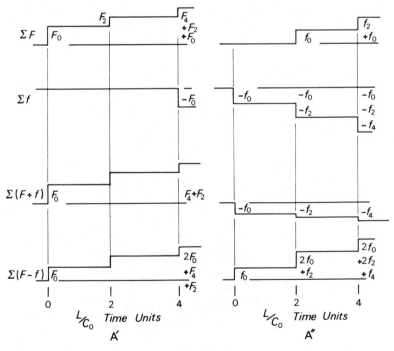

Fig. 6.3

This rigorous solution will obviously lead to a very involved graphical solution. However, writing Equation (6.6) as follows:

$$h_{A_2'} - h_{A_2''} = h_{-0} \frac{-C_0}{g} \left[1 + \frac{a_1}{a_2} \right] \left[V_{A_2'} - V_P + V_P - V_{A_{-0}''} \right] - 2f_0$$

$$= h_0 \frac{-C_0}{g} \left[1 + \frac{a_1}{a_2} \right] \left[V_{A_2'} - V_P \right] + 2F_0 \qquad (6.7)$$

shows that the construction characteristic passes through P and thus $W_1 P$ can be deleted to shorten the procedure.

The next and most general solution is involved in finding A_4. Application of Equation (6.4) and reference to Fig. 6.3 shows that:

$$h_{A_4'} - h_{A_4''} = h_{-0} \frac{-C_0}{g} \left[1 + \frac{a_1}{a_2} \right] \left[V_{A_4'} - V_{A_{-0}''} \right] - 2F_0 - 2f_0 - 2f_2 \quad (6.8)$$

Rearranging the above:

$$h_{A_4''} = h_{-0} \frac{-C_0}{g} \left[1 + \frac{a_1}{a_2} \right] \left[V_{A_4'} - V_Q + V_Q - V_{A_{-0}''} \right] - 2F_0 - 2f_0 - 2f_2$$

$$= h_{-0} \frac{-C_0}{g} \left[1 + \frac{a_1}{a_2} \right] \left[V_{A_4'} - V_Q \right] - 2F_0 + 2F_2 \qquad (6.9)$$

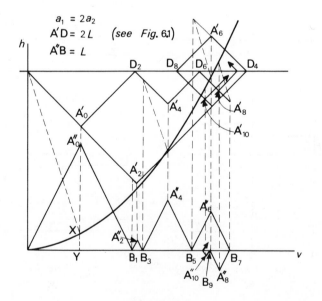

Fig. 6.4

Thus the construction characteristic passes through Q since it can be shown that QW_2 is equal to $-2F_0 + 2F_2$. Q is on the characteristic $D_2 A_4'$ directly above B_3 as in the previous case.

The example just illustrated assumed that the transient head difference across the gate at all times would conform to the gate characteristic, e.g. $Z_4 X_4 = A_4' A_4''$. The same characteristic would apply in the system if steady flow existed, but as friction is neglected, the conditions would necessarily be adjusted until h_0 only occurred across the gate.

6.2   Problems

6.2.1   Illustrate the graphical solution for an instantaneous gate opening.

Fig. 6.4 contains the solution for this problem. The general case has areas of a_1 and a_2 but in this solution a convenient ratio of $a_1/a_2 = 2$ has been used. The slope of the propagation characteristic intersecting the gate characteristic in this case is therefore $3C_0/g$.

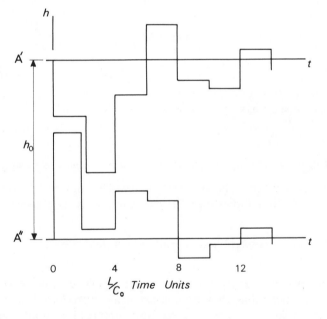

Fig. 6.5

In Fig. 6.5 the pressure time traces for A' and A" have been drawn. The pressure heads approach the respective reservoir levels since the assumption in this example that h_{-0} is dropped across the gate must apply because this

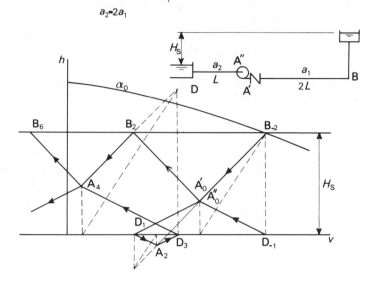

$$V_{A'} = V_{A'} = V_{A''} = V_{A''} \frac{a_2}{a_1}$$

$$a_2 = 2a_1$$

Fig. 6.6

is the limit of the definition of the gate characteristic at this stage. It is also clear in this solution how the gate characteristic stipulates that the head (XY in Fig. 6.4) must always equal that between A' and A" (XY = A'₀A"₀) under transient conditions.

6.2.2   Illustrate the graphical solution for an instantaneous pump failure.

In Fig. 6.6 the solution is shown for a system with suction and delivery pipes of unequal length and diameter. Friction and inertia have been ignored.

6.2.3   Illustrate the solution for a pump start including suction and delivery pipes.

In Fig. 6.7 the case for equal diameters and pipe lengths is worked. It is important to realise that a pump connected to a line produces low pressures on the suction side. As will be discussed later, rupture of the water column, if it occurs, may eventually produce significant overpressure on the suction side. Thus the connection of pump boosters, e.g. for fire protection, (see Sharp and Coulson, 1968) to a pipeline which is part of a gravity flow system might produce disastrous effects because of overpressures that were never anticipated at the design stage.

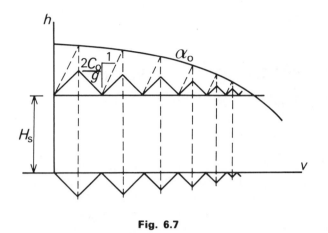

Fig. 6.7

6.2.4   Indicate the requirements for computer analysis with propagations both upstream and downstream.

A control which is responsible for disturbances which must propagate in both directions will require special provisions in a computer analysis. The graphical solutions treated above draw a physical picture of the events and also indicate that the treatment of the problem numerically can be achieved by the principle that:

$$\Delta h = \pm \frac{C_0}{g}\left(1 + \frac{a_1}{a_2}\right)\Delta v \qquad (6.10)$$

the sign depending on whether the control is regarded as upstream or downstream.

The consequences of this is that the velocity changes on each side of the control, which must satisfy a continuity relation, produce pressure head changes, which when added become Δh. This would, of course, be in addition to any pressure head that the control itself demands for the residual velocity through it.

7

Compounding – a change in diameter (a junction)

7.1 Introduction

In Chapter 6 the propagation of effects in different diameters required a continuity relation, Equation (6.1). However, the change in diameter is not always at the source of primary water hammer and so propagations can be regarded as approaching the junction from one direction.

A principle of profound importance which emerges and which is relevent not only to junctions but to any consideration where velocity changes can occur during transmission must be fully understood. In passing through the junction there will be a velocity change, hence there will be a change in water hammer pressures which, as outlined in Chapter 6, must propagate equally upstream and downstream. Thus at the junction there will be a reflection, or reverse transmission, the amount depending on the extent of the velocity change. It should be noted that such a reflection together with a moderation of the transmission in the forward direction will also occur if there is any change in the propagation characteristic (C_0/g). In these introductory comments secondary losses are ignored.

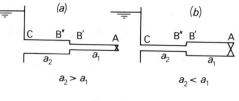

Fig. 7.1

The junction of two pipes can have two orientations for a primary transmission. In Fig. 7.1 the two alternatives are shown. Both require the continuity relation:

$$v_{B'} a_1 = v_{B''} a_2 \tag{7.1}$$

where B′ and B″ are downstream and upstream of B respectively. Velocities may be scaled as before so that

$$V_{B'} = v_{B'} \quad \text{and} \quad V_{B''} = v_{B''} \frac{a_2}{a_1}$$

Suppose that C_0/g is the same for each pipe then in AB propagations will occur using characteristic slopes $\pm C_0/g$ whereas in BC the slopes will be

$$\pm \frac{C_0 \, a_1}{g \, a_2}.$$

The wave propagations may be described as follows:

$$h_{B'_{-1}} - h_{A_0} = \frac{-C_0}{g}\left(V_{B'_{-1}} - V_{A_0}\right) \qquad (7.2)$$

$$h_{C_0} - h_{B''_1} = \frac{-C_0}{g}\left(v_{C_0} - v_{B''_1}\right)$$

$$= \frac{-C_0 \, a_1}{g \, a_2}\left(V_{C_0} - V_{B''_1}\right) \qquad (7.3)$$

for the cases of an observer arriving at A_0 and B''_1 respectively.

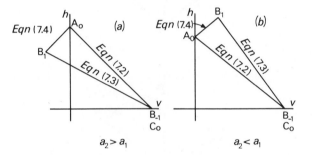

Fig. 7.2

To illustrate the principles the case of a gate closing instantaneously at A_0 will be considered. A primary F propagation will result which will arrive at B at time $t = 1$ and thus

$$h_{A_0} - h_{B_1} = \frac{C_0}{g}\left(V_{A_0} - V_{B_1}\right) \qquad (7.4)$$

The combination of Equations (7.3) and (7.4) provide the solution for the conditions at B_1 (see Fig. 7.2) and since the scaling of velocities is known, the velocity corresponding to B' or B" may be recovered. The head must be the same on either side of B and so throughout the remainder of the analysis B' and B" will be dispensed with.

Continuing the solution for a system ABC which has $a_2 = 1.30 \, a_1$, $L_{AB} = L_{BC}$, and C_0/g the same for both AB and BC, in Fig. 7.3, the transmissions and reflections are apparent, developing into a complex interaction of propagations. In a sense the junction is continually generating propagations in both directions.

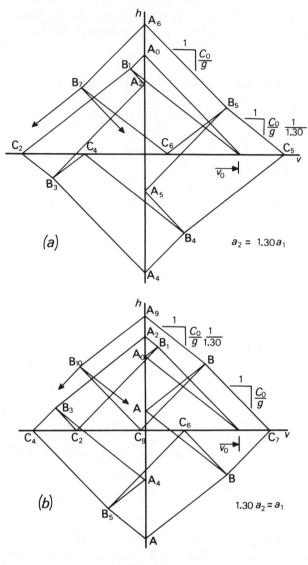

Fig. 7.3

7.2   Problems

7.2.1   Does the orientation of pipes in series affect the amount of water hammer?

To make a comparison of effects in any system with alternatives requires not only a common basis such as the effectiveness of the system, discharge

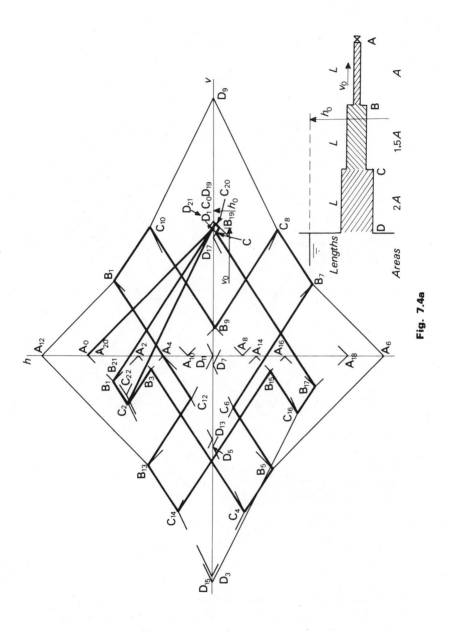

Fig. 7.4a

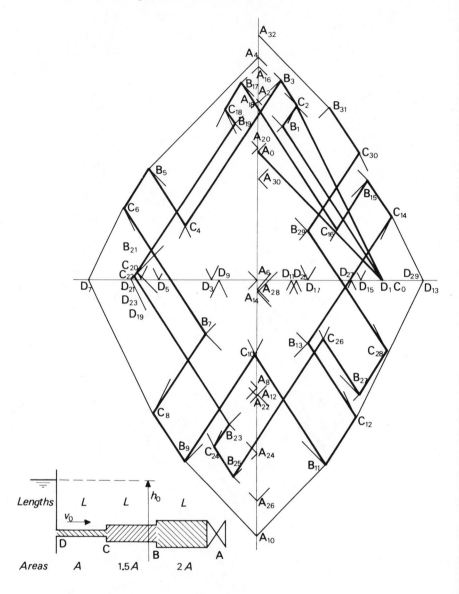

Fig. 7.4b

and head, but, in the case of water hammer, time is a criterion. Thus in a system of three pipes in series which delivers the same discharge between two reservoirs with any orientation of the pipes, a downstream valve closure can be considered as a design water hammer problem.

However, the pressure head time record can differ so vastly that comparisons can only be justified if there is a known cycle of water hammer effects so that the history of all possible events was realised in each case.

It is possible to concoct a system of three pipes in series which has a finite cyclic solution in time, when friction is ignored, and one such example has been analysed graphically for two configurations as shown in Fig. 7.4a and b, for the extreme case of an instantaneous downstream gate closure. The pressure head time records at various points are shown in Fig. 7.5 and Fig. 7.6 respectively and events are essentially cyclic as required, and indicated. The flow rate through the two systems is identical when the velocity in AB in Fig. 7.4a is twice that in Fig. 7.4b. The pressure heads can then be compared and it is found that in the case of (b) with diverging pipes the water hammer is 88 % of case (a) with converging pipes in a time less than $10 L/C_0$, whereas in a longer time the amount is only 70 %.

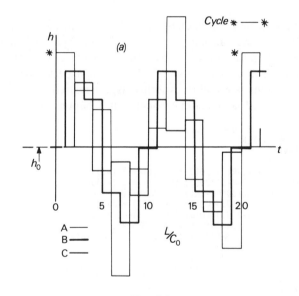

Fig. 7.5

Therefore case (b) is superior from the point of view of water hammer and could lead to savings in pipe material. From the steady flow point of view the pipe orientation is not normal as one would expect the arrangement of pipes to converge in the gravity flow direction. This study suggests that what is suitable for the design considerations for steady operational flows may not give the best orientation of the system for the design dynamic loads.

Propagations in the respective pipes will then require characteristic slopes as follows for a constant C_0/g:

$$\text{AB, } \pm\frac{C_0}{g}: \text{BC, } \pm\frac{a_1}{a_2}\frac{C_0}{g}: \text{BD, } \pm\frac{a_1}{a_3}\frac{C_0}{g}$$

There is the additional requirement that $h_B = h_{B'} = h_{B''}$ at the junction at all times.

The propagation equations require a strict sign convention and thus it is necessary to decide which direction will be used for positive velocity in each pipe and then the directions of the F and f propagations are determined. The directions used in this introductory example have been shown in Fig. 8.1.

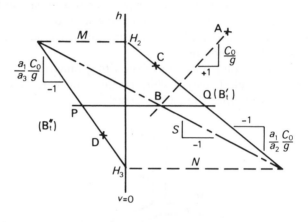

Fig. 8.2

In principle, the method of solution is to combine the effect of two pipes having similar propagations into one equivalent characteristic. In Fig. 8.2 it is assumed that conditions from C and D are being propagated to reach B simultaneously. The convention of positive velocity in Fig. 8.1 then requires characteristic negative slopes for both pipes as these will be f propagations. A trapezium is completed where these characteristics cut the $v = 0$ axis of the h–v diagram. If the intersections on the $v = 0$ axis are at H_2 and H_3 and the horizontal parallel sides are of magnitude M and N, then the slope of the diagonal S is expressed by the equation:

$$S = \frac{H_2 - H_3}{M + N} \tag{8.4}$$

Writing the magnitudes M and N as

$$V_D - \frac{a_3}{a_1}\frac{g}{C_0}(H_2 - h_D) \quad \text{and} \quad V_C - \frac{a_2}{a_1}\frac{g}{C_0}(h_C - H_3)$$

respectively, then

$$S = \frac{H_2 - H_3}{-\dfrac{a_3}{a_1}\dfrac{g}{C_0}(h_D - H_3 + H_2 - h_D) - \dfrac{a_2}{a_1}\dfrac{g}{C_0}(H_2 - h_C + h_C - H_3)}$$

$$= \frac{-1}{\dfrac{a_3}{a_1}\dfrac{g}{C_0} + \dfrac{a_2}{a_1}\dfrac{g}{C_0}} \tag{8.5}$$

Thus the slope of the trapezium is independent of the locations of C and D and is only a function of the pipeline constants.

Suppose now that an F propagation arrives from A at B at the same instant as those from C and D and let it intersect the trapezium diagonal at B. A horizontal line ($h = $ constant) from B cuts the other characteristics at P and Q and the velocities of these points are respectively V_B, V_P and V_Q. Consideration of the geometry of the triangles establishes the following:

$$V_Q = \frac{H_2 - h_B}{H_2 - H_3}(N)$$

$$V_P = \frac{h_B - H_3}{H_2 - H_3}(-M)$$

$$V_B = \frac{H_2 - h_B}{H_2 - H_3}(M + N) - M$$

$$= \frac{H_2 N - h_B N - h_B M + H_3 M}{H_2 - H_3} = V_Q + V_P \tag{8.6}$$

It is evident therefore that the intersections B, Q and P satisfy the simultaneous requirements of flow continuity and head compatibility at B and if the propagations simultaneously arrive at time t at B then these points may be designated B_t, B'_t and B''_t respectively.

This is the essence of the construction and any problem first requires the determination of the slope S either by construction, i.e. complete the trapezium, or by calculation using an equation such as (8.5). The slope S is constant for any problem but its location is determined by the finding of either of the corners of the trapezium.

Numerical solutions using the method of characteristics reflect a similar principle as can be seen in Equation (A3.23) in Appendix 3, which gives the new pressure head at the junction and uses a multiplier not unlike Equation (8.5).

8.2   Problems

8.2.1   In graphical solutions how does the sign convention of velocity affect the construction?

Referring to Fig. 8.3, the trapezium is formed by using pipes which have the same sign for the propagation of effects to the junction and hence the slope of the diagonal changes. However, as the convention must be predetermined, the construction will not alter as the solution proceeds.

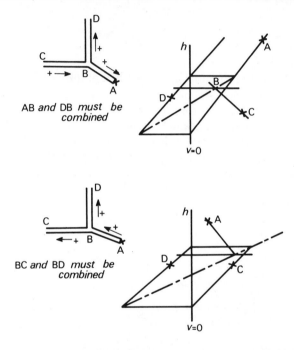

Fig. 8.3

8.2.2   Solve graphically the case of an instantaneous valve closure on a main line having one branch line.

In Fig. 8.4 the three pipes meet at junction B and the area ratios are as shown. For convenience, to demonstrate the solution, the pipes are of equal length and C_0/g is the same in each pipe. Since the two reservoirs are at the same level, flow would normally occur from each, but in the simplifying treatment of no friction it can be specified that there is no flow from D. The velocity convention of velocity positive toward B in BC and BD provides the first trapezium RSTU and the diagonal RT. After the initial water hammer effect at A_0 the first event at B_1 is found with a dash-dot line

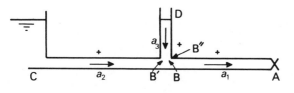

$a_3 : a_2 : a_1 = 3 : 1.5 : 1$

$AB = BC = BD$

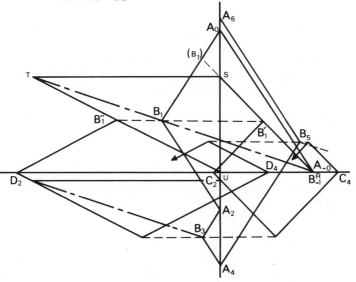

Fig. 8.4

representing the diagonal intersected by the $+ C_0/g$ propagation from A_0. Reflections and transmissions then take place automatically.

It is clear that the water hammer in AB would be reduced in passing through to BC because of the change in area, but the tee has a much more dramatic effect as (B_1) falls to B_1 (see Fig. 8.4). This is the principle of the open surge tank. The remaining solution takes no account of possible changes in water level in the reservoir D which would occur if it had a finite volume.

8.2.3   Compare the branch solution for water hammer with the rigid column solution of an open surge tank.

The rigid column, or mass surge, theory uses the dynamic flow equation (2.4) but assumes that the water column CBD in Fig. 8.5 is instantly aware

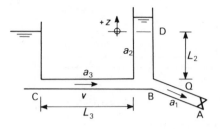

Fig. 8.5

that the force associated with its acceleration is the difference in pressure head at its extremes, i.e. z (positive upwards). Allowing for the total mass being accelerated in BC and BD the equation becomes:

$$\left(\frac{L'}{g}\right)\frac{dv}{dt} = -z \tag{8.7}$$

where $L' = L_3 + \dfrac{a_3}{a_2}L_2$.

A continuity equation is necessary to account for conditions at the junction B, as follows:

$$va_3 = a_2\frac{dz}{dt} + Q \tag{8.8}$$

In the case when Q is reduced instantaneously to zero, and ignoring friction and velocity head, the solution of the above equations is a sinusoidal oscillation:

$$z = z_m \sin 2\pi\frac{t}{T} \tag{8.9}$$

where the maximum value of z is given by:

$$z_m = \frac{Q}{a_1}\left(L'\frac{a_3}{g\,a_2}\right)^{\frac{1}{2}} \tag{8.10}$$

and the period of the oscillation is:

$$T = 2\pi\left(L'\frac{a_2}{g\,a_3}\right)^{\frac{1}{2}} \tag{8.11}$$

Consider now an example with the following data: $L_1 = L_2 = 76.2$ m (250 ft), $L_3 = 304.8$ m (1000 ft); $a_2 = 2a_1 = 2a_3 = 0.186$ m² (2 ft²); $C_0/g = 100$; and $Q = 0.0283$ m³ s⁻¹ (1 ft³ s⁻¹). In Fig. 8.6 part of the graphical solution of the water hammer is indicated and obviously the presence of the

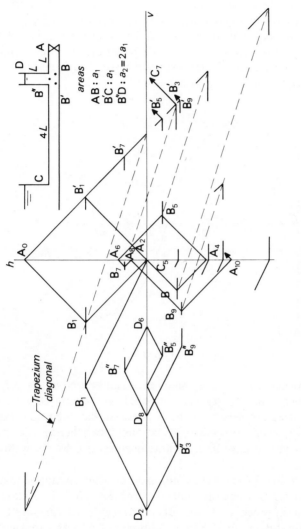

areas

AB : a_1
B'C : a_1
B"D : $a_2 = 2a_1$

Trapezium diagonal

Fig. 8.6

branch BD reduces the pressure head in AB from point A_0 to the level B_1 in each pipe. There will be many oscillations and even without friction an allowance should be made for the variation of the water level at D as the water in BD develops velocity.

The complete solution in this case is more readily and more accurately obtained using a computer and the result is plotted in Fig. 8.7 where the level variation of D with time is shown. This shows that z_m is 1.14 m (3.75 ft) in the first oscillation and T is 52.5 s. Employing the same data and using the equations derived from the mass surge analysis, Equations (8.10) and (8.11), yields $z_m = 1.27$ m (4.18 ft) and $T = 52.4$ s.

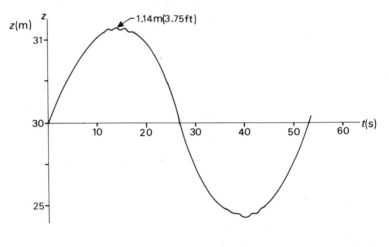

Fig. 8.7

The agreement is not exact because, as can be seen in Fig. 8.7, there are high frequency oscillations superimposed on the long period mass surge and these need not necessarily be in phase so that one would expect z_m to vary slightly with successive oscillations. Nevertheless, the mass surge solution seems to have been accurately confirmed by the water hammer analysis.

However, Fig. 8.6 shows how misleading the interpretation that the mass surge analysis is, therefore, sufficient can be. Thus, it is clear that the pressure in the pipelines BC and BD (at B) is at least as great as 15 m (50 ft) above static at B_1. In Fig. 8.8 the conditions at B are compared with those at D for approximately $T/4$. These extreme oscillations of pressure at B are, of course, propagated into BC.

In this simplified example it is evident that the mass surge analysis alone does not develop a complete awareness of the problem of pipeline protection. In practice, such an extreme example might not arise but with

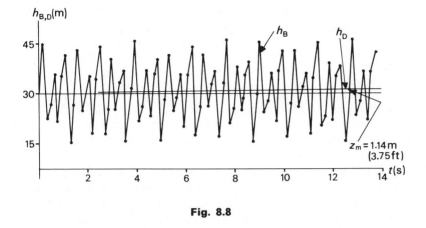

Fig. 8.8

modern computational methods it is certain that a complete water hammer analysis should always be possible and advantageous.

8.2.4 How is the branch problem solved graphically when there is more than one pipe branching from the same point?

The solution for a cross, for example, is solved in exactly the same way as was illustrated in Problem 8.2.2. The trapezium diagonal is found by progressive pairing of pipes. Thus two pipes would combine to form one

```
        SUBROUTINE BRANCH(..........)
        DO 10 J=1,NP-1
        J1=J+1
        I=N(J)+1
        M=N(L)+1
        VR = V(J,I)-R(J)*CO(J)*(V(J,I)-V(J,I-1))
        HR = H(J,I)-R(J)*CO(J)*(H(J,I)-H(J,I-1))
        VRR=VR-G/CO(J)*HR-FK(J)*VR*ABS(VR)
        VS=V(J1,1)-R(J1)*CO(J1)*(V(J1,1)-V(J1,2))
        HS=H(J1,1)-R(J1)*CO(J1)*(H(J1,1)-H(J1,2))
        VSS=VS+G/CO(J+1)*HS-FK(J+1)*VS*ABS(VS)
        VIR=V(L,M)-R(L)*CO(L)*(V(L,M)-V(L,M-1))
        HIR=H(L,M)-R(L)*CO(L)*(H(L,M)-H(L,M-1))
        VIRR=VIR-G/CO(L)*HIR-FK(L)*VIR*ABS(VIR)
        HP(J,I)=(VRR*AR(J)-VSS*AR(J1)+VIRR*AR(L))/(G/CO(J)*AR(J)+G/CO(J1)*
       1AR(J1)+G/CO(L)*AR(L))*(-1)
        HP(J1,1)=HP(J,I)
        HP(L,M)=HP(J,I)
        VP(J,I)=VRR+G/CO(J)*HP(J,I)
        VP(J1,1)=VSS-G/CO(J1)*HP(J,I)
        VP(L,M)=VIRR+G/CO(L)*HP(J,I)
   10   CONTINUE
        RETURN
        END
```

Fig. P.7

equivalent propagation which is then combined with the next pipe, of similar sign, to form the ultimate propagation representing three pipes, and so on.

8.2.5   Develop a computer solution for a branch pipeline.

In Appendix 3 the finite difference equations for a BRANCH are indicated. The convention adopted here is a positive velocity away from the main line and thus propagations in the J pipe and the L (Branch) pipe are the same kind as they approach the branch. Thus the grid points are $N(J)+1$ and $N(L)+1$ at the branch in pipe J and L respectively. In the SUBROUTINE BRANCH (Fig. P.7), the key statements necessary to provide the solution are included. For NP mainline pipes there could only be $NP-1$ branches. Of course there are now more than NP pipes in the total system which must be DIMENSIONed and the housekeeping is increased accordingly.

9
Pseudo-friction

9.1 Introduction

Friction losses in pipes in unsteady flow have been allowed for in various ways mostly by assuming that it is concentrated at certain points (Parmakian, 1963). Consider, for example, the sudden gate closure shown in Fig. 9.1. The hydraulic friction loss under a steady state flow of v_0 is h_f and would place A_{-0} as shown. The gate closure requires the propagation from B_{-1} which, to provide the proper reference conditions, would originate from the same point as A_{-0}. The succeeding construction has all the

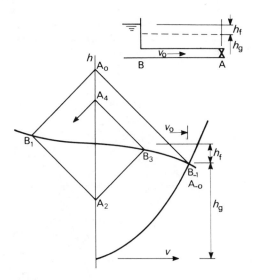

Fig. 9.1

appearances of providing the desired attenuation of the water hammer effects with time. This method due to Schnyder (see Jaeger, 1956) is often referred to as allowing for friction, but it is in fact providing the attenuation by an orifice loss, the orifice being situated just to the right of B.

65

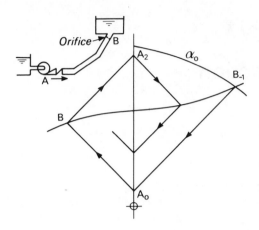

Fig. 9.2

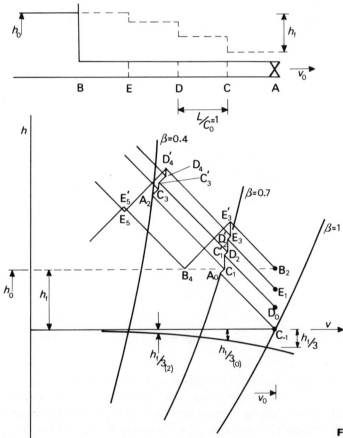

Fig. 9.3

9.2   Problems

9.2.1   Show the solution for an instantaneous pump failure allowing for a pseudo-friction loss.

In Fig. 9.2, the pump failure case is considered (ignoring inertia) allowing for an orifice loss at B to represent the effect of friction. A non-return valve is assumed to function and it is apparent that the water hammer attenuates with time.

9.2.2   Simulate friction loss with a distribution of orifices and solve the gate gradual closure case graphically.

In Fig. 9.3 a pipeline is assumed to have three orifices uniformly distributed along it at C, D and E. Conditions upstream of these constrictions are represented by a prime. The amount of friction loss required at each velocity is stepped off the head loss parabola, but a trial and error process is involved until the sum of the losses equals the total friction at a representative velocity spanning the calculations made for the three orifices.

It is worth noting that the solution gives the appearance of a reflection from each orifice but it is basically a change of datum from which a propagation is transmitted when determining the next event. Thus to find A_2 when a new gate characteristic ($\beta = 0.4$) is specified, conditions are propagated from C_1 which is displaced due to the fact that a new velocity has been produced corresponding to the first gate movement ($\beta = 0.7$).

10

Rational friction

10.1 Introduction

The solution of the basic water hammer equations and application to the graphical method omitted the friction term in Equation (2.4) and reference to this was made in Problem 2.2.2. The rationalization of the effect of friction has been lucidly described by Gibson (1945) and its incorporation in the graphical method was made possible by Gerny (1949), but unfortunately little publicity was given to this until a discussion by Sharp (1974) of a paper by Wood and Jones (1973).

The method to be described is called the Gibson–Gerny method and its implementation should be no more difficult or laborious than the pseudo-representation in Chapter 9. It has the advantage that it presents a physically correct picture of friction, in so far as any of the methods are true when based on steady state friction behaviour.

The losses due to friction are properly called distributed losses as opposed to lumped losses, e.g. at a valve, and are to be considered later.

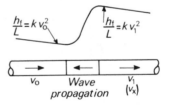

Fig. 10.1

In Fig. 10.1 a velocity v_0 is required to be changed to v_1 by the passage of a wave. If the friction obeys the quadratic law:

$$h_f = kLv^2 \tag{10.1}$$

then the head which must be absorbed corresponds to the change in friction head:

$$\delta h = \pm k\delta x \,(v_0{}^2 \pm v_x{}^2) \tag{10.2}$$

in which v_x is the final velocity (instead of v_1) due to the change in friction

68

effects in moving a distance δx. The unbalanced pressure head δh must be propagated equally in all directions and therefore it is distributed in this case equally upstream and downstream.

The additional velocity change which will be propagated with the wave is then:

$$\delta v = 1/2 \, c_1 \frac{g}{C_0} \, (v_0{}^2 \pm v_x{}^2) \, \delta x \tag{10.3}$$

Physically, therefore, $1/2 \, \delta h$ is the attenuation of the advancing wave and an equal and opposite amount is propagated in the opposite direction.

If the wave following an instantaneous gate closure takes L/C_0 time units to reach a reservoir at the upstream end, the reflection seen at the gate will continue to arrive until $2L/C_0$ and hence the phenomenon known as 'head recovery' will be a feature of friction effects as seen by an observer who remains at the gate. Thus a square wave takes on a 'sawtooth' appearance as shown in Fig. 10.2.

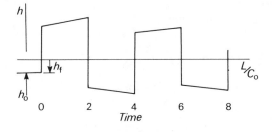

Fig. 10.2

Equation (10.3) has three important solutions depending on the relative signs of v_0 and v_x and whether or not $v_0 = 0$.

(1) $v_0 = 0$

$$\frac{dv}{dx} = -\frac{kg}{2C_0} v_x{}^2$$

$$v_x - v_1 = -v_1 \left[\frac{1}{1 + \dfrac{2C_0}{gkv_1 x}} \right] \tag{10.4}$$

(2) v_0 and v_1 the same sign

$$\frac{dv}{dx} = \frac{kg}{2C_0} \left[v_0{}^2 - v_x{}^2 \right]$$

$$v_x - v_1 = (v_0 + v_1) \frac{e^{Nx} - 1}{M e^{Nx} + 1} \tag{10.5}$$

where $M = \dfrac{v_0 + v_1}{v_0 - v_1}$ and $N = \dfrac{kg}{C_0} v_0$

(3) v_0 and v_1 of opposite sign

$$\frac{dv}{dx} = \frac{kg}{2C_0}(v_0{}^2 + v_x{}^2)$$

$$v_x - v_1 = \frac{v_0{}^2 + v_1{}^2}{v_0 \cot(1/2\,Nx) - v_1} \tag{10.6}$$

For the three cases above where v_x lies between v_0 and v_1 a good approximation is:

$$v_x - v_1 = \frac{v_0{}^2 - pv_1{}^2}{r + pv_1} \tag{10.7}$$

where $r = \dfrac{2C_0}{kgx}$ and $p = 1$ and has the same sign as v_1/v_0. Equation (10.7) can be rearranged in the form:

$$(v_x - v_1)\frac{2C_0}{g} = kxv_0{}^2 - kxpv_1{}^2 \tag{10.8}$$

since friction is usually much less than $200\,v_1$.

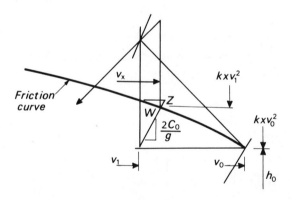

Fig. 10.3

In Fig. 10.3 the application of this equation to the graphical construction is illustrated. The value of v_1 is known and the change $(v_x - v_1)$ is then $g/(2C_0)$ times the friction change. There is negligible error in the use of W instead of Z for finding v_x.

10.2   Problems

10.2.1   Illustrate the Gibson–Gerny rationalization of friction in terms of the basic Equations (2.26) and (2.27).

The basic equations referred to are in terms of F and f type propagations. The additional effects $\frac{1}{2}\delta h$ related to the change δv in Equation (10.3) are modifications of the F and f propagations, with regard to their pressure head significance, which in the ideal no friction case were unchanged in form. The graphical method can readily incorporate these changes by adding and subtracting the developing head, and velocity, changes due to friction which are linear with time.

In Fig. 10.4 the addition of the effects of friction are taken in the summations $\Sigma(F+f)$ and $\Sigma(F-f)$, which are required in Equations (2.24) and (2.25) respectively. As an illustration the orthodox problem of a gate closure at A is considered in steps of $2L/C_0$, i.e. 2 time units, in a time greater than 8 for convenience in setting out the solution.

In Fig. 10.4a the point A_0 is on the first gate closure characteristic and v_x is found by using the $2C_0/g$ construction for a friction change of δh_0. The friction head changes for a total length of $4L$ are shown. The change δh_0 is required in $4L/C_0$ units of time of which half will be a reduction of F_0 as it passes from A to B and the other half will appear as a propagation in the opposite direction BA, i.e. an f propagation.

An observer at A sees the conditions summarized in Fig. 10.4b. Thus, at the gate A at time 8 the total effect of $\delta h_0/2$ has reached there as well as the addition of $\delta h_4/2$ which has accompanied the returning f propagation due to the reflection of the modified F_0 at B.

By inspection of the geometry in Fig. 10.4a it is confirmed that the characteristic F propagation along a slope $-C_0/g$ to find A_8 (on the next gate characteristic) passes through W and the required water hammer equation:

$$
\begin{aligned}
h_{A_8} &= h_0 + 2\Sigma f - \frac{C_0}{g}\left(v_{A_8} - v_0\right) \\
&= h_0 - 2(F_0 - \delta h_0) + \delta h_4 - \frac{C_0}{g}\left(v_{A_8} - v_0\right)
\end{aligned}
\tag{10.9}
$$

is satisfied. The construction, therefore, provides the solution for conditions at A, the control end, directly. In order to recover other information the following interpretation is used.

In Fig. 10.5 the division of the change in friction δh_0 into two parts is shown as the return wave which appears as a head recovery at the control end and the modification (a reduction) of the advancing wave. Thus, A_0 rises to A_{-2}, which means just before A_2, and the propagation towards B is

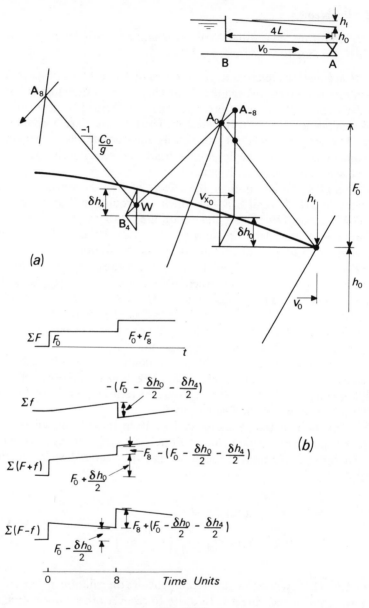

Fig. 10.4

through A_0 since the modified wave when reaching B has undergone a datum change of δh_0.

Similarly, B_1 reaches B_{-3} through a combination of a return wave and a

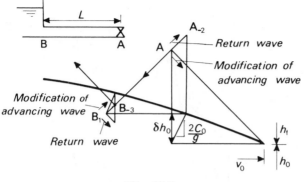

Fig. 10.5

datum change and the propagation passes through B_{-3} which is equivalent to the point W in Fig. 10.4. There is, therefore, a rational interpretation of the changing conditions for anywhere in the system using this method for friction. It remains to consider a more general case with superposition of other waves and an extension of the problem in Fig. 10.4 is used for this purpose.

In Fig. 10.6 another gate closure position in addition to that shown in Fig. 10.4 is considered. At A_2 the further closure adds an F_2 as shown in Fig. 10.7 which includes all the F and f propagations seen by an observer who remains at A. In this case, waves pass each other at D, which is a distance L from B, and they become influenced by the change in velocity remaining which produces different friction changes. A wave passing from A_2 to B_6 must propagate a return from D_5 towards A_8 amounting to $3/4$ $\delta h'_5/2$ by the time it reaches A_8. This is an f propagation and is positive in sign. Simultaneously, the f propagation passing from B_4 to A_8 is modified after reaching D_5 an amount of $-3/4\,\delta h''_5/2$ as it progresses and arrives at A_8. The total f propagations due to the original major changes at A become at time $8L/C_0$:

$$-\left(F_0 - \frac{\delta h_0}{2} - \frac{1}{4}\frac{\delta h_4}{2} - \frac{3}{4}\frac{\delta h'_5}{2} + \frac{3}{4}\frac{\delta h''_5}{2}\right) = f_8 \qquad (10.10)$$

When the additional f propagations which comprise return effects are included it is found that:

$$h_{A_8} = h_0 + 2(f_8) + \frac{\delta h_0}{2} + \frac{3}{4}\frac{(\delta h_2 - \delta h_0)}{2} - \frac{C_0}{g}\left(v_{A_8} - v_0\right) \qquad (10.11)$$

In the construction it may be observed that:

$$\delta h'_5 + \delta h_2 = \delta h''_5 + \delta h_4 + \delta h_0 \qquad (10.12)$$

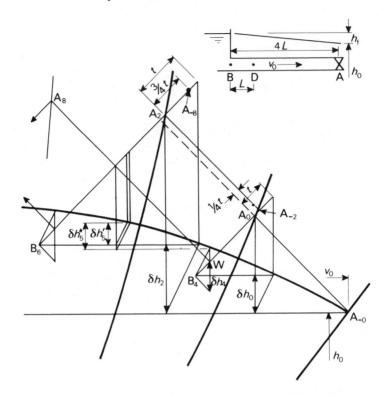

Fig. 10.6

Thus, with rearrangement and substitution of Equation (10.12) in Equation (10.11),

$$h_{A_8} = h_0 - 2(F_0 - \delta h_0) + \delta h_4 - \frac{C_0}{g}\left(v_{A_8} - v_0\right) \tag{10.13}$$

It is proved, therefore, that the addition of the F_2 propagation does not affect the location of A_8 although the conditions within the system are now much more complex.

To complete the description it is readily shown that:

$$h_{A_2} = h_0 + \frac{1}{2}\frac{\delta h_0}{2} - \frac{C_0}{g}\left(v_{A_2} - v_0\right) \tag{10.14}$$

which means that the location of A_2 is through A_{-2} as shown in Fig. 10.6. Conditions will change from A_2 to A_{-8} unless any further gate closure occurs. In this construction it is clear there would be only slight error in carrying the propagation through A_0 as shown by the dotted line and using the last $3/4t$ instead of the first as shown in the full line construction.

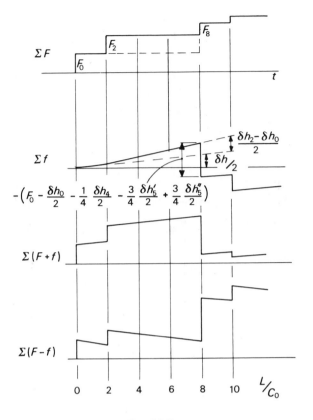

Fig. 10.7

10.2.2   What further implications can be derived from the 'sawtooth' pressure head record referred to on p. 69?

The pressure head record shown in Fig. 10.2 would be seen by an observer at the gate. It is one of the few relatively straightforward field results that might be expected and could be used to confirm the rationalization of friction. The record must in fact include the effects of imperfect reflections at the reservoir and the gate (fully closed), but these may be small in a pipeline with significant steady state friction. Supposing the latter situation to be true, the graphical solution would appear as shown in Fig. 10.8. From the geometry:

$$\frac{y_0}{y_1} = \frac{F_0 - ab/2}{F_0 - (ab + bc + cd/2)} \qquad (10.15)$$

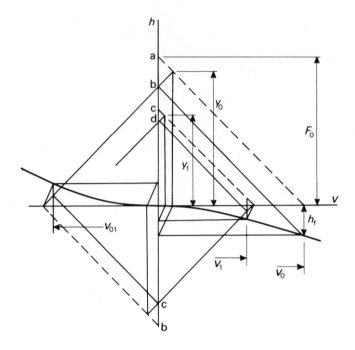

Fig. 10.8

$$= \frac{1 - \dfrac{1}{2}\dfrac{kLv_0^2}{F_0}}{1 - \dfrac{kL}{2F_0}(2v_0^2 + 2v_{01}^2 + v_1^2)} \qquad (10.16)$$

which may be approximated using the rules for small increments beginning with $\gamma_1 = ab + bc + \dfrac{cd}{2} \ll F_0$ so that $(1 - \gamma_1)^{-1} = 1 + \gamma_1$ etc.

$$\frac{y_0}{y_1} = 1 + \frac{kL}{2F_0}(-v_0^2 + 2v_0^2 + 2v_{01}^2 + v_1^2) \qquad (10.17)$$

A further approximation using:

$$v_{01} = v_0 - \Delta = v_1 + \Delta$$

enables Equation (10.17) to be expressed as:

$$\frac{y_0}{y_1} = 1 + \frac{kL}{F_0}(v_0^2 + v_1^2) \qquad (10.18)$$

If the decay of the pressure at A is exponential then:

$$\frac{y_0}{y_1} = \frac{\gamma e^{-\alpha t_0}}{\gamma e^{-\alpha t_1}} = e^{\alpha(t_1-t_0)} = e^{4L\alpha/C_0}$$

$$= 1 + \frac{4L}{C_0}\alpha \text{ (approximately)} \tag{10.19}$$

Thus, in the above derivation, the supposition (for the ideal case of an instantaneous gate closure) that the decrease in pressure at the gate is due to friction leads to the approximate relation:

$$\alpha = \text{(average friction per unit length per unit velocity) } g/2 \tag{10.20}$$

An approach such as this may be used to evaluate under controlled laboratory conditions the effects of friction and reflection since the exponential decay at the gate is relatively easily measured.

10.2.3   Is the inclusion of friction in the numerical solution by the method of characteristics physically superior to the Gibson–Gerny method?

The method of characteristics solved by the finite difference scheme in Appendix 3 included a number of linearizations. Firstly, the determination of the conditions in the x, t plane for points inside those specified at the grid points, e.g. A and B, as shown in Fig. A3.1 is a linear approximation. Secondly, the friction terms in Equations (A3.3) and (A3.4) are only g/C_0 times the friction and as such are merely expressing that the head and velocity changes are related as in Equation (2.30), as would be expected. Thus, the numerical solution is the same in principle as the Gibson–Gerny interpretation.

The numerical methods, however, perform the task of keeping track of all events automatically. At present all methods suffer from the weakness that they are expressible in terms of quasi-steady friction which is not the time-dependent friction that is demanded when water hammer is present.

Some work has been done on this complex phenomenon (e.g. Van Emmerik, 1964; Wood, 1966; Zielke, 1968) and much more is needed since the mechanism of friction attenuation of pressure during water hammer plays a very important part in highly damped systems, e.g. oil pipelines and plastic pipelines. It will become evident in the subsequent discussion of rupture of the water column that some features of that phenomenon can be elucidated only by the existence of friction in the system. Although neglect of friction in calculations usually yields conservative answers its proper evaluation will become more important if other matters are to be rationalized fully.

11
Non-linear gate

11.1 Introduction

In Chapter 4, the gate characteristic, as a boundary condition, was dealt with in the classical manner expounded by Allievi. In this text the word *linear* has been used to describe a gate that is consistent with studies of transients in the absence of friction such that for any steady state position:

$$v = \beta v_0 \tag{11.1}$$

A *uniform* gate operation on the other hand requires β to be linear with time.

It will become clear that the gate characteristic under real conditions, i.e. when fluid friction is present, is a *non-linear* problem and a method will be developed which approximates real conditions satisfactorily and which has been confirmed by laboratory tests. In this context we are referring to the functional relationship rather than the method of gate operation.

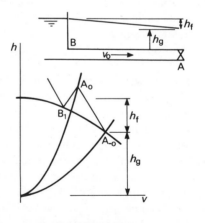

Fig. 11.1

In Fig. 11.1 there appears to be a little difficulty in describing graphically the friction characteristic (h_f), two gate characteristic parabolas associated with a gate closure and the propagation characteristic $A_{-0}A_0$, A_0B_1, etc. The majority of texts, however, rather loosely define h_g as possibly being

78

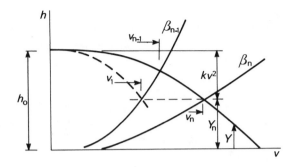

Fig. 11.2

the nozzle loss as the gate discharges to atmosphere. It certainly cannot be the loss across a gate that is connected downstream at A to a large reservoir. One would normally expect a well-designed valve to have negligible head loss across it when it is full open. If this was true, then an equation of the form:

$$v = \beta v_0 \left(\frac{h}{h_0}\right)^{\frac{1}{2}}$$ (Equation 4.3)

cannot provide the customary definition of β, since $h = 0$, when $v = v_0$. By allowing a very small loss across the gate when it is fully open (see Crowe, 1968; Sharp, 1969), Equation (4.3) is written with an additional term:

$$v_n = \beta_n K Y_n^{\frac{1}{2}}$$ (11.2)

where v_n and Y_n are defined in Fig. 11.2 and:

$$K = v_0/(h_0 - k v_0^2)^{\frac{1}{2}}$$ (11.3)

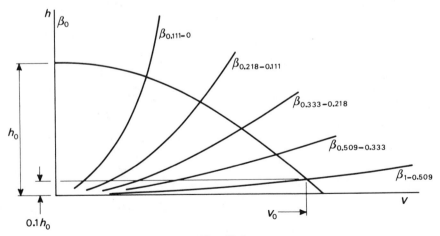

Fig. 11.3

When the gate is fully open a finite value of K is determined by setting $h_0 - kv^2$ at a convenient small value such as Ph_0, where $P = 0.03$ for example. Hence, when $Y_n = 0.1 h_0$, $v_n = v_0$ and $\beta = 1$, and when $Y_n = h_0$, $v_n = 0$ since $\beta = 0$. The extreme values of β remain as in the simple linear solution.

Inspection of Fig. 11.3, however, shows that the variation of β, and hence gate opening, differs markedly from the classical solution illustrated in Fig. 4.2. β has been taken as the ratio of the area of the gate opening (see Equation 4.4) and the example shows there is very little reduction of velocity until β is less than 0.5.

11.2 Problems

11.2.1 Compare the non-linear and linear (no friction) results for a uniform gate closure.

The comparison is made in the Fig. 11.4 taking the linear result Fig. 4.2b for identical values of h_0, v_0 and time of total closure. It was assumed that $\beta \propto (T-t)$, i.e. a uniform gate closure. For the purposes of this comparison T is taken as much less than $2L/C_0$ so that head recovery, due to pipe friction, is not considered.

It is evident that the very last part of the gate closure (approximately 10%) is that which produces significant water hammer and the elementary gate characteristic produces a very misleading result. In this example of a uniform gate closure the pressure head rises very sharply as the gate nears its fully closed position. In the case of gate closures in greater than $2L/C_0$ it is

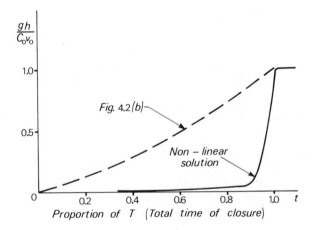

Fig. 11.4

clear that the early part of the gate closure could be accomplished quickly, whereas the bulk of the time of closure should be confined to the last 10% of gate movement. This is generally confirmed by field experience.

11.2.2 Discuss the relationship between gate behaviour and gate position.

Various gate types and the associated water hammer have been reported (Sharp, 1969; Bernhart, 1976; Parmakian, 1963) and valve manufacturers have indicated the steady state flow characteristics of particular types such as the butterfly valve (MacLellan and Carruthers, 1976). To define the gate characteristic for a water hammer problem requires the net flow area, local energy losses and physical gate position in relation to the pipeline flow

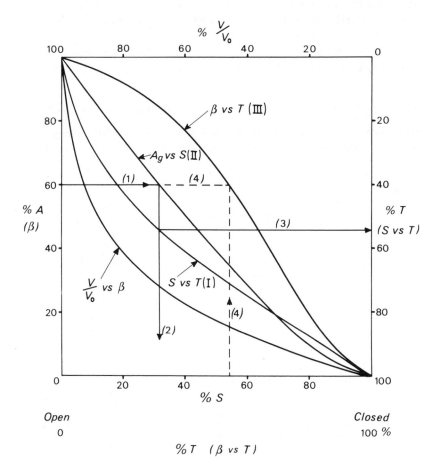

Fig. 11.5

velocity to be determined. This is quite complex and will require some assumptions to be made. The information shown in Fig. 11.5 is an example. Suppose the physical gate position in time is represented by curve I, which is the stroke (S) versus time (T) (see Sharp, 1969). The gate is a rotary (plug-type) valve and, using the projected area as a measure of gate opening (see Parmakian, 1963), the curve II can be calculated, which represents the area of the gate (A_g) versus S.

It is possible to calculate a coefficient of discharge (C_D) based on steady flow which is unrealistic when water hammer is present and so, rather than do this, it is assumed $C_D = 1$ and the parameter β can then be calculated using curves I and II and producing curve III. Paths (1), (2), (3) and (4) are followed as shown. The non-linear dimensionless variation of β with v/v_0 depicted in Fig. 11.3 is now used and is plotted for reference in Fig. 11.5. The data (curve III) enables values of time to be assigned to the gate characteristic shown in Fig. 11.3.

The above procedure was followed in making comparisons with a laboratory experiment, shown in Fig. 11.6. Good agreement is evident.

11.2.3 Is there an optimum gate closure for times greater than $2L/C_0$?

In the case of a linear gate, work has been done (Cabelka and Franc, 1959) for the optimum (least) water hammer for T_0 greater than $2L/C_0$. Their

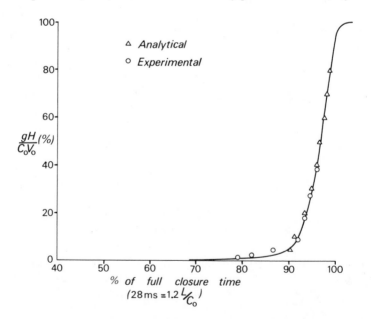

Fig. 11.6

result is of historical interest since it had provided an early guide for gate operation. It may be summarized by the equation:

$$\frac{h'}{h_0} = \frac{C_0}{g} \frac{v_0}{h_0} \frac{1}{(T_c - 1)} \tag{11.4}$$

where h' is the optimum pressure head during gate closure, i.e. the least water hammer that must be incurred. It should be noted that Bernard Michel (Cabelka and Franc, 1959) showed that the above was merely a special case for the linear gate.

As an illustration of the optimum result and referring to Fig. 11.7, reflections modify the continuing process of gate closure after time $T = 2$. The linear gate requires the variation of β (for $h = $ constant) to be linear

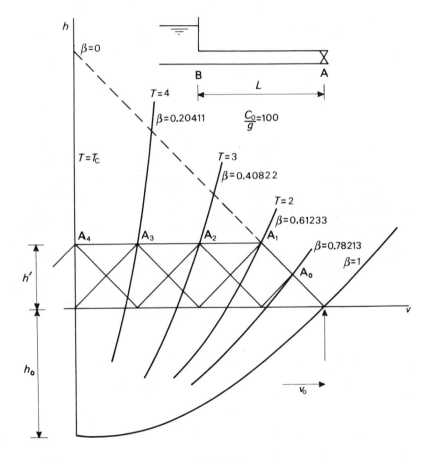

Fig. 11.7

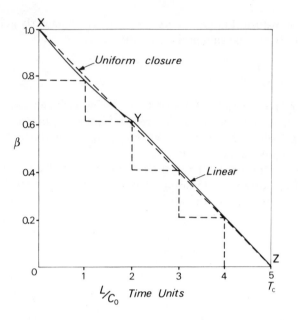

Fig. 11.8

with velocity and so the variation of β with time as shown in Fig. 11.8 will be linear between Y and Z. However, as pointed out in Section 4.2 and Fig. 4.2b, the portion XY in Fig. 11.8 corresponding to the gate operation up to $2L/C_0$ is not a straight line. The optimum closure in this example, in a time of $T_c = 5L/C_0$ in five steps, shows that the solution on the assumption of a linear gate yields a β/T relation that is nearly uniform as pointed out by Cabelka and Franc.

For the non-linear solution developed in this text an example is worked by the graphical analysis as shown in Fig. 11.9 to show how the solution is achieved using the rational Gibson–Gerny method for friction. The result was obtained by a trial and error in the choice of the optimum value of HNH.

11.2.4   Develop a computer program for the optimum gate closure case.

The solution of the optimum gate closure as illustrated above is easily achieved with the computer for those cases where T_c is an integer. Thus using the Program OPT shown in Fig. P.8 a solution is obtained based on the graphical solution directly. This program yields the values of β at integer times T, e.g. 2, 4, and 6 in Fig. 11.9.

In the Program OPT (Fig. P.8), the values of G (gravity), CO (wave

```
      PROGRAM OPT
C     **  TOL=SMALL POSITIVE VALUE  **
      COG=CO/G
      VO=Q/AR
      FK=HSTAT*(1.0-P)/VO**2
      BK=VO/SQRT(P*HSTAT)
      IT=ITC/2
8     CONTINUE
      VEL=-VO
      VEL2=-VO**2
10    HNH=HNH+0.001
      CONS=HNH+(1.-P)*HSTAT
      CALL QUAD (FK,COG,VEL,VEL2,CONS,VQ)
      VV(1)=VQ
      FKO=FK*(VO**2-VV(1)**2)/2.
      V(1)=VV(1)-FKO/COG
      H(1)=HSTAT+HNH-FKO
      B(1)=V(1)/(BK*SQRT(H(1)))
      VEL=-VV(1)
      VEL2=VV(1)**2
      CONS=2.*(HNH+HSTAT*(1.-P)-FK*VO**2)
      DO 20 I=2,IT
      CALL QUAD (FK,+2.*COG,VEL,VEL2,CONS,VQ)
      VB(I-1)=VQ
      HB(I-1)=H(I-1)-COG*(V(I-1)-VB(I-1))
      VEL=-VQ
      VEL2=-VQ**2
      CONS=HNH+HSTAT-HB(I-1)
      CALL QUAD (FK,COG,VEL,VEL2,CONS,VQ)
      VV(I)=VQ
      V(I)=VB(I-1)+(VV(I)-VV(I-1))/2.
      H(I)=HB(I-1)+COG*(VB(I-1)-V(I))
      B(I)=V(I)/(BK*SQRT(H(I)))
      VEL=-VV(I)
      VEL2=VV(I)**2
      CONS=2.*(HSTAT+HNH-HB(I-1)-FK*(VV(I-1)**2+VB(I-1)**2)/2.)
20    CONTINUE
      IF(V(IT).GT.TOL)GO TO 8
      CALL EXIT
      END



      SUBROUTINE QUAD (A,B,VEL,VEL2,CONS,VQ)
      Y=CONS+B*VEL+A*VEL2
      VQ=B*(SQRT(1.0-4.0*A*Y/B**2)-1.0)/(2.0*A)
      RETURN
      END
```

Fig. P.8

speed), Q (discharge), AR (area), P (initial loss constant, see p. 80), TOL (residual error when gate fully closed) and HSTAT (head across the gate when fully closed) are INPUT. An initial HNH, the optimum head is assumed and incremented by 0.001 until the final velocity is within the TOL for a given ITC (time of closure in L/C_0 units).

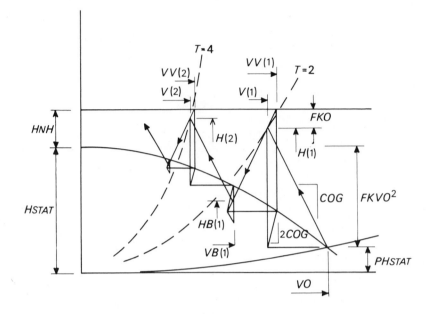

Fig. 11.9

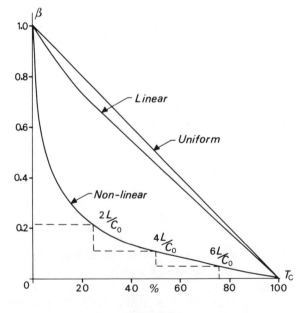

Fig. 11.10

The program determines all the information associated with its graphical counterpart as shown in Fig. 11.9. When the values of β, B(I) in the program, are plotted as in Fig. 11.10 it is clear that any path may be traced, e.g. the dotted line, to connect these points and the same optimum result would occur. In practice, however, a continuous curve would be used and the analysis of a number of cases suggests that a suitable closing rate for optimum pressure surge is an exponential curve.

11.2.5   Compare the various optimum gate closure cases.

The graphical analysis is an approximation, even with friction by the rational method and it is convenient to use the computer for numerical solutions based on the method of characteristics. The β–T values from the exponential curve shown in Fig. 11.10 were used and compared with the graphical solution in Fig. 11.11. In this result the deviation from the value of HNH = 11 m occurs because it is very sensitive to the shape of the curve up to $2L/C_0$ but in general the exponential assumption gives a fair result.

The comparison is now extended to the solutions for a linear gate for the same v_0, h_0, C_0, L and T_c as follows:

	Optimum pressure head, h' (m)
De Sparre formula	20.3
Linear gate	17.1
Non-linear gate (graphical)	11.0
Non-linear gate (exponential using computer)	14.6
Linear gate with friction	81.5°

° The values of β and T of the linear gate optimum solution were used for the exact analysis of a single pipeline with friction according to the method of characteristics and yielded this alarming result.

11.2.6   Is there a generalized optimum gate closure solution?

Assuming that an exponential gate closure provides the optimum pressure head, the way that the optimum head (h') varies with time of closure and initial velocity can be determined using a computer analysis. Alternatively, the graphical solution (Program OPT), may be used and by this means the generalized solution has been established, as shown in the dimensionless plot in Fig. 11.12. It is clear from this result how the time of closure must be chosen so that, for a given initial velocity, and wave speed, the minimum water hammer that must occur can be determined and realized, provided that the gate closure is approximately exponential.

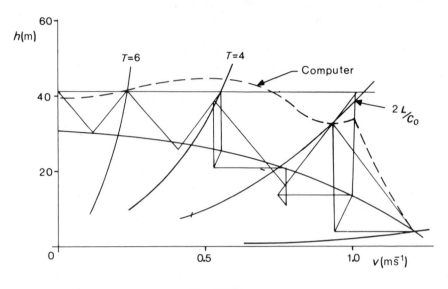

Fig. 11.11

11.2.7   Calculate the water hammer due to a gradual valve closure.

For a continuous closure of a gate the water hammer will be a function of the initial velocity and the total time and rate of closure. A valve may be a vertical gate, a butterfly type or a ball type etc. Each will have certain characteristic losses which may not be appropriate to the dynamic conditions during water hammer and the rate of closure is mostly unregulated. The preceding examples enable us only to specify the limits, i.e. the maximum or minimum water hammer if the time of closure is the sole criterion.

For example, a pipeline 8 km long, 600 mm diameter, discharges $600 \, \mathrm{l \, s^{-1}}$ of wastewater with 60 m fall and must occasionally be shut down by a downstream gate control. What time of closure is required to limit the water hammer to 20 m? The velocity is $2.12 \, \mathrm{m \, s^{-1}}$ and the maximum water hammer is 212 m (wave speed $= 980 \, \mathrm{m \, s^{-1}}$), so a very rapid closure is out of the question. If the gate is closed in the best possible fashion, $h' = 20$ m and N (see Fig. 11.12) will be $2.1 \times 2.12/0.6 = 7.42$ and thus the time which must be exceeded is $7.42 \times 7.14 = 53$ s. If the gate closure is other than optimal this might need to be increased sixfold.

11.2.8   Develop a program for a downstream valve.

In Appendix 3, the downstream gate valve equations are expressed for the non-linear solution. As indicated, it is necessary to assume a value for Ph_0,

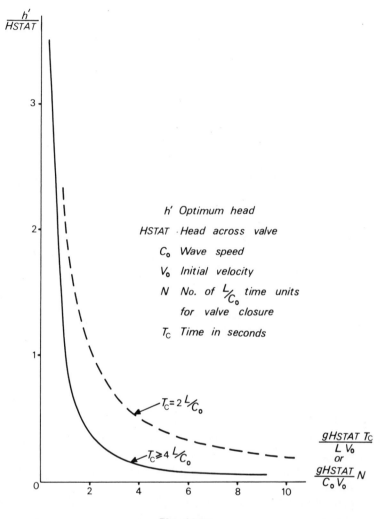

$\dfrac{h'}{HSTAT}$

h' Optimum head

HSTAT · Head across valve

C_0 Wave speed

V_0 Initial velocity

N No. of L/C_0 time units
for valve closure

T_C Time in seconds

$T_C = 2 \, L/C_0$

$T_C \geqslant 4 \, L/C_0$

$\dfrac{gHSTAT \, T_C}{L \, V_0}$
or
$\dfrac{gHSTAT}{C_0 \, V_0} N$

Fig. 11.12

the head loss across the valve when it is fully open (PP = 0.03). In the analysis of complex systems, it has to be decided what is in effect the rated head across the valve when it is fully closed (called HEC). The valve when it is fully open then allows this amount of pressure head for friction in the pipeline(s) associated with it.

The SUBROUTINE VALVE (Fig. P.9) is a straightforward application of the finite difference equations, but it must be stressed that the boundary conditions are all important when deciding the operation which will be

```
SUBROUTINE VALVE(.............)
PP=0.03
J=1
I=1
CON=V(J,I)**2*CO(J)/(HEC*G)
VS = V(J,I)-R(J)*CO(J)*(V(J,I)-V(J,I+1))
HS = H(J,I)-R(J)*CO(J)*(H(J,I)-H(J,I+1))
VSS=VS+G/CO(J)*HS-FK(J)*VS*ABS(VS)
B=1.
IF(MM.LE.M1) GO TO 10
IF(MM.GE.M2) GO TO 15
A1=MM-M1
A2=M2-M1
B=EXP(-A2/A1)
GO TO 30
10      CONTINUE
HP(J,I)=HDELI
VP(J,I)=VS-G/CO(J)*(HP(J,I)-HS)-FK(J)*VS*ABS(VS)
GO TO 40
15      CONTINUE
B=0.
VP(J,I)=0.
HP(J,I)=(VSS-VP(J,I))*CO(J)/G
GO TO 40
30      CONTINUE
IF(V(J,I).GT.0.) GO TO 35
CON=-CON
35      CONTINUE
RAD=1.+4.*PP*(VSS-G/CO(J)*HDELI)/(D**2*CON)
IF(RAD.LE.0.) GO TO 15
VP(J,I)=-B**2*CON/(2.*PP)*(1.-SORT(RAD))
HP(J,I)=(VSS-VP(J,I))*CO(J)/G
40      CONTINUE
RETURN
END
```

Fig. P.9

responsible for water hammer. The SUBROUTINE includes an exponential type movement to correspond with the conclusions about non-linear gate behaviour. This, however, is only true if the valve external motion is apparently uniform with time.

The constant HDELI is the delivery level of the valve and when the time (MM) is less than M1, the valve discharges freely to this level and if the time is greater than M2 the valve is fully closed.

12

Rupture of the water column

12.1 Introduction

Rupture or separation of the water column occurs frequently in pipelines and some of the physical behaviour can be deduced with the help of the graphical analysis. Unlike some aspects of system behaviour that are conservative when neglected, this phenomenon in general increases the problems and severity of water hammer. The phenomenon, rupture of the water column, is the discontinuity in the liquid created by excessive tension when pressures are lowered near to vapour pressure. Liquids may sustain very great tension (Harvey *et al.*, 1947) when special care, e.g. removal of foreign matter, is taken but in practical systems such ideal conditions do not exist.

It must be assumed that there is a host of nuclei in the liquid which may contain holes with entrapped air or else the pipe wall itself is rough, even though hydraulically smooth, and thus nucleation centres are available for liquid to vaporize when the pressure drops near to vapour pressure. For the purposes of analytical treatment, the critical value of pressure necessary to cause the 'cavitation' rupture will be assumed as the vapour pressure for the prevailing ambient temperature conditions.

In previous analysis there was no limit to the fall of pressure that was permissible and hence the most important property of superposition of waves was utilized. The pressure head may not be made dimensionless, however, when rupture is considered and absolute pressure must be retained, as has been the practice throughout this text. As in all preliminary analysis friction is neglected initially, although the role it plays in determining the complexity of the problem is in fact most important as subsequent details will show.

In Fig. 12.1 a horizontal pipeline is considered with a downstream gate at A and an upstream reservoir at B. An instantaneous closure produces A_0 and B_1 proceeds as usual for this simple case. Suppose now that the vapour pressure limit is included and an observer moves with an f propagation from B_1 to arrive at A at time 2. Instead of a zero velocity and pressure head corresponding to A_2' (due to gate closed at A) the pressure is unable to fall beyond A_2. This suggests a residual velocity v_r of the liquid at A away from the gate. As the liquid vaporizes at these pressures it is usually assumed that

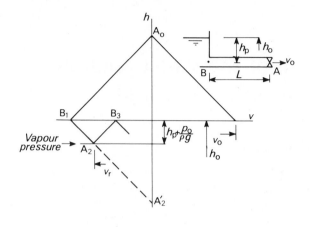

Fig. 12.1

a cavity of water vapour remains to fill the space left by the receding plane front of liquid.

The analysis proceeds on the assumption that the vapour space may accommodate any desired velocity as long as the pressure remains at vapour pressure. As a 'steady' velocity of the liquid occurs in between each change it is clear that a summation of the product of these velocities by the time interval yields a length, which may be taken as the length of the cavity that grows or collapses accordingly as the velocity is negative or positive respectively at A. Thus the cavity due to the rupture of the water column in the first instance is due to a combined effect of no velocity permitted at the gate and pressures near vapour pressure. The size and life of the cavity is a function of the residual velocity at A and the time.

In practice such ideal conditions as an horizontal pipeline seldom exist and thus in Fig. 12.2a the more useful condition of a knee in a rising main is shown. In this instance the pressure head may fall to A_0 due to pump failure as shown in Fig. 12.2b, i.e. to suction tank level, but the pressure may only fall as low as D for the pipe at the knee due to the limitation of $h_p + p_0/\rho g$ below the hydraulic gradient.

It appears that rupture of the water column is induced at D at time L_1/C_0 and after this it is anticipated that separate oscillations of pressure in AD and BD take place while the discontinuity (cavity) exists at D. Eventually the water column reunites, often with devastating effect and the analysis of this problem therefore becomes one of the most important in system design.

As the pipeline elevation now enters the problem, rather more difficulties are faced in performing an analysis. In order to grasp the main factors which control the growth and collapse of cavities it will be convenient to consider the case of a horizontal pipeline.

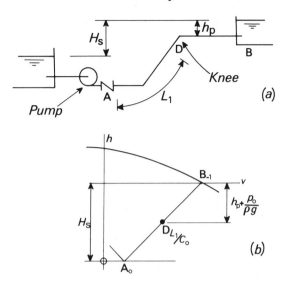

Fig. 12.2

12.2 Problems

12.2.1   Develop a simple model for rupture with the graphical analysis (ignoring friction).

In Fig. 12.3 a solution for rupture of the water column is shown and corresponds to the type of analysis described by Lupton (1953) and it will be shown that the cavities may form not only at the gate A but also at intermediate points due to the interaction of waves. The process is described in some detail as follows.

At A_2 the residual velocity in the water column which flows from A to B (leaving a cavity at A), persists for a time $2L/C_0$. In the velocity time insert (b) these residual velocities are shown. At time A_4 the residual velocity suddenly becomes positive. Pressures at A are still at vapour pressure and the shaded area in (b) represents the volume of cavity per unit area which has been formed. The positive excursions of velocity at A must provide eventually a volume per unit area which equals the size of cavity and at this instant it will disappear. The time this occurs can be calculated as:

$$0.857 \times 2 = 0.429 \times 2 + 1.715 \times T$$

and hence, $T = 0.5$ and the time of closure is then $t = 6.5$ s. At time 6 an F propagation leaves A and proceeds towards B_7. An f propagation from B at time 5.5 arrives at $A_{6.5}$ and finds the closed gate since at this instant the cavity has disappeared. Hence $A_{6.5}$ occurs on the $v = 0$ axis and an

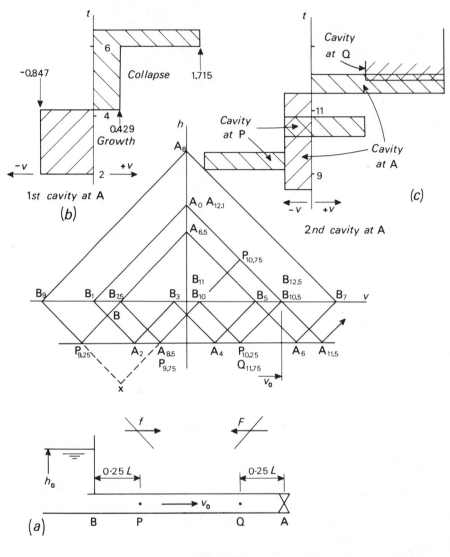

Fig. 12.3

additional F propagation proceeds towards $B_{7.5}$. These two propagations could proceed independently and any result found by superposition but for the fact that nowhere in the system is the pressure allowed to fall below vapour pressure.

Firstly, following $B_{7.5}$ to $A_{8.5}$ another cavity is formed at A and the water column moves towards B with the residual velocity corresponding to $A_{8.5}$.

Meanwhile, an f propagation from B_9 would encounter the former at a point P at time 9.25. The superposition of these two effects would produce point X in Fig. 12.3 which is impossible and it must be assumed a cavity forms at P. The water column between A and P will continue to move with the velocity corresponding to $A_{8.5}$ whereas that between P and B will move towards B with the greater velocity corresponding to $P_{9.25}$. The intermediate cavity at P grows with a velocity corresponding to the difference of those at $P_{9.25}$ and $A_{8.5}$ as shown in the insert (c). The cavity at P grows and collapses before any effect can be transmitted to A since it is assumed that the cavity constitutes a discontinuity in the flow.

In this analysis a complementary cavity occurs subsequently at Q and the remaining events become most complicated, because with each collapse of a cavity at a time not corresponding to a multiple of $2L/C_0$ time units, new pressure waves are created which interact with others with the possibility of another discontinuity.

Not all texts recognize the interaction of the pressure waves (Bergeron, 1961) and in some cases, including the example worked in Fig. 12.3, the action of friction in attenuating the waves would reduce the pressure changes so that the vapour pressure is no longer reached and thus there is a limit to the growing complexity due to the production of new waves. There is no doubt, however, about the occurrence of the intermediate cavities as these have been observed in laboratory tests (Sharp, 1965).

12.2.2   Develop a simple model including friction.

When friction is considered, a serious criticism of the previous analysis must arise which invalidates all but the notion of the first cavity. The role that friction plays can be satisfied by the simple physical assumption of an energy loss gradient in the positive velocity direction. Consider the water column after time 2 in Fig. 12.3a. It is assumed that it moves with the residual velocity corresponding to A_2 away from the gate. The upstream end of the flowing water column is at vapour pressure (a cavity at A) and if friction exists it would be expected that an hydraulic gradient corresponding to the friction loss would require a decrease in pressure head in the direction of flow. This situation is impossible with an horizontal pipeline, as the pressure is already at a minimum for the system. Thus the analysis cannot proceed beyond A_2, as suggested above, as some mechanism is lacking to explain the dynamics of flow (Sharp, 1965, 1966). The analysis of rupture, therefore, requires additional concepts since even this relatively straightforward case is unable to proceed any further.

One approach has concentrated on the equilibrium of a single vapour cavity which might, in its initial growing stage, be regarded as spherical. Thus the equilibrium of a cavity due to surface tension (σ) (Johnson, 1963),

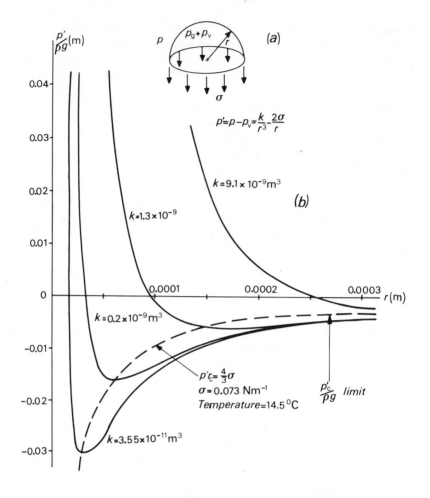

Fig. 12.4

as shown in Fig. 12.4, is divided into two distinct zones. In one zone, above the dashed critical dividing line, the cavity is in stable equilibrium. In the other zone, below the dividing line, an increase in p', the ambient pressure, relative to vapour pressure, is associated with an increase in the cavity size. This latter zone is, therefore, a condition of unstable equilibrium and suggests that the cavity can increase in size as required as the external pressure head increases.

Referring now to Fig. 12.5a, the water column can move away with a residual velocity v_r in a horizontal pipeline, since the associated hydraulic gradient for friction is allowed to develop as the cavity increases in size.

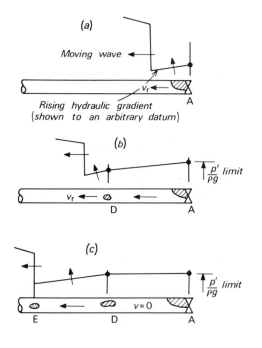

Fig. 12.5

Ultimately, however, a limit will be reached, e.g. $p'/(\rho g)$, which forces further cavity formation at regular distances, as shown in Fig. 12.5b and c. An important associated effect is that the velocity in between the cavities goes to zero.

This simple model has prompted a simulation (Sharp, 1977), which works reasonably well with a gradually sloping pipeline where regularly spaced cavities are generated when the pressure head falls to vapour pressure. It should be noted that these theories suggest that cavities may have different origins, some are due to wave interaction while others are possibly forced as in the last development discussed above.

Some researchers consider that the phenomenon should be further divided into two types, separation with free surface flow and cavitating or bubbly flow (Kranenburg, 1974). In the former there is a significant part of the flow where there is no longer continuity of the water column and free surface wave effects would require an entirely different treatment. In the latter type there might be a satisfactory approximation by using a lower wave speed for the bubbly mixture as the emphasis of more recent research has tended to show (Tullis *et al.*, 1976; Martin *et al.*, 1976; Beattie, 1976; see also Enever, 1972).

Studies have been made of the effects of gas release (Dijkman and

Vreugdenhil, 1969; Kranenburg, 1972; Swaffield, 1972) which can be of significant concern when volatile liquids are involved. Most studies have been devoted to horizontal pipelines and mostly in controlled laboratory situations (Li, 1962; Siemons, 1967; Kalkwijk and Kranenburg, 1971; see also Duc, 1965).

12.3 Summary

In summary, the problem of water column rupture is most complex, and the logical approach for solving engineering problems would seem to be to determine with the best analytical methods the possibility of rupture and its extent if it is to occur, and, in other than the most simple of pipelines which might be analysed for rupture, prevent it from occurring if possible. In present day practice the majority of pipe system studies are preoccupied with the low pressure problem and this is affecting the class of pipe used and the design viewpoint for protection measures.

13
Protection methods

13.1 Introduction

The use of some form of protection device is largely a matter of economics. It is physically possible to provide a pipeline capable of withstanding any surge or shock, so that failure does not occur. However, it is more reasonable to design a pipeline for pressures only slightly in excess of normal steady operating values. If occasional large shocks must be provided for, it will generally cost less to prevent them from reaching the majority of the pipe system by some form of protective device.

Unfortunately, in many cases, proper consideration of water hammer only occurs after something calamitous has happened. In this event true economy can no longer be satisfactorily achieved and the form of protection may be seriously restricted. If a system has been constructed that is prone to bad water hammer because of inadequate design or unusual site conditions, considerable expertise is necessary to solve the problem.

Two types of problem will be encountered. One type is the effect of an abrupt shock, e.g. the failure of an electrically powered pump, which is unavoidable. The other is the resonance problem which arises because the system amplifies oscillations as described in Chapter 14. The latter is of particular interest for the continuous control of a system such as hydroelectric or oil hydraulic. The impedance method is a direct way of studying the resonance features and for modifications to a system the method is basically similar to that used for electrical circuits where judicious use is made of capacitance and inductance to change the circuit impedance.

This chapter is devoted to a brief description of the significant features of commonly used methods for protection against abrupt shocks due to water hammer. The graphical analysis is the most lucid method for demonstrating many of the basic principles.

13.2 Problems

13.2.1 How can the flywheel effect be utilized?

The rotating inertia of a machine is the flywheel effect and is clearly associated with the start up or run down of a pump or turbine. As the water hammer is a function of the rapidity of the change of velocity one way of providing protection, for small installations, is to increase the flywheel effect.

The method has much to recommend it since there is no ancillary equipment or moving parts subject to wear. The required GD^2 is easily found and the water hammer calculated according to the methods of Chapter 5. The torque requirements for starting will be altered however and in the case of electric motors the type of starter may have to be changed.

13.2.2   When is the open surge tank appropriate?

The open surge tank has been illustrated in Chapter 8. Its simplicity secures its application in those instances where an installation near the basic machine (pump or turbine) is feasible. Hydroelectric installations employ an open surge tank quite frequently although the stability problems in the case of small turbine load fluctuations and cost may require more sophisticated forms such as the differential surge tank (see Jaeger, 1956). In pumping schemes it is seldom possible to install an open surge tank on the delivery side and closed surge tanks (see p. 101) are used, whereas in the case of protection for a suction main they may be feasible.

An example of the surge tank has already been given and it remains to emphasize the need to ensure that the connection to the pipeline is as generous as possible, unless throttling is envisaged, otherwise rather more water hammer passes the surge tank than simple theory implies.

In present day practice more attention is being paid to the security of supply to essential industries. If there is a low pressure in a system there is a tendency (and temptation) to add booster pumps, but such an action changes the dynamic conditions in the low pressure pipeline. One such example was the case of a large thermal power station complex using fossil fuel (Sharp and Coulson, 1968). A continuous supply of fresh water was required for fire protection, cooling and other operational needs and when boosting out of a low pressure supply main was attempted unexpected water hammer occurred causing a major water supply failure. The protection measure adopted (open surge tanks) was influenced significantly by the realization that the breakdown of the water supply threatened the electric power supply to a large number of essential consumers.

13.2.3   Describe the air vessel solution for a pumping main.

In the case of pumping plant it is usually impractical to construct an open surge tank and so it is customary to provide a closed air vessel when some other form of flow control, e.g. valve surge suppressor, is not used as the primary method of protection. With a sealed air volume the flow of liquid in and out of the tank, controlled by pressure differences, will have to be compatible with the behaviour of the air space. It is customary to assume a relationship of the form

$$pV^n = \text{constant} \tag{13.1}$$

where p and V are the absolute pressure and volume, respectively, of the air space.

Field tests have suggested values for n which is normally taken as constant. Laboratory tests (Graze, 1968) confirmed that n varies significantly throughout a cycle of volume fluctuation and that an exact thermodynamic relation should be used. However, in the absence of more specific data regarding heat transfer coefficients $n = 1.3$ is a commonly assumed value for prototype behaviour. In the field a number of complicating features, e.g. the type of pipe connection, imprecise operating data, the actual role of associated control valves and unknown heat transfer coefficients, introduce uncertainties that could exceed the contribution due to the selection of a constant n.

When the analysis is set up for computer solution, however, it is a formality to conduct sensitivity studies and a satisfactory economical engineering solution can usually be found, whichever analytical method is used. The choice of air volume is basically trial and error but a first estimate can be obtained by using charts (Parmakian, 1963) which requires a decision to be made concerning the maximum and minimum values of absolute pressure permitted adjacent to the pump.

An example based on the graphical analysis is illustrated in Fig. 13.1 for the case of a power failure to a pumping plant. The pipeline is 26 km long, diameter 0.76 m with $C_0 = 1000$ m.s^{-1}. At the upstream end at the pump flange the air vessel has an initial volume of 36 m^3. The static lift is 114 m with 11.6 m friction and $Q_0 = 0.326$ m^3 s^{-1}. Each point is found by trial and error such that the air vessel volume change equals the mean pipeline discharge in the time interval L/C_0 while the absolute pressure head in the air vessel found from the pipeline pressure head satisfies $H_{\text{abs}} V^{1.3}$ = constant. The pressure head time result at the upstream end is shown in Fig. 13.1. The solution is worked for a large time interval so that the general construction can be easily followed.

The ancillary equipment associated with an air vessel must be considered carefully. Usually a compressor is required with air lines to the upper part of the air vessel. Depending on the exposure of the air vessel to ambient temperatures, various degrees of automatic regulation of the water level in the air vessel and alarm devices will be required. The reliability of the system rests heavily on such ancillary equipment. If the air vessel provides absolute protection, i.e. without it pipeline failure is a certainty, then it might be

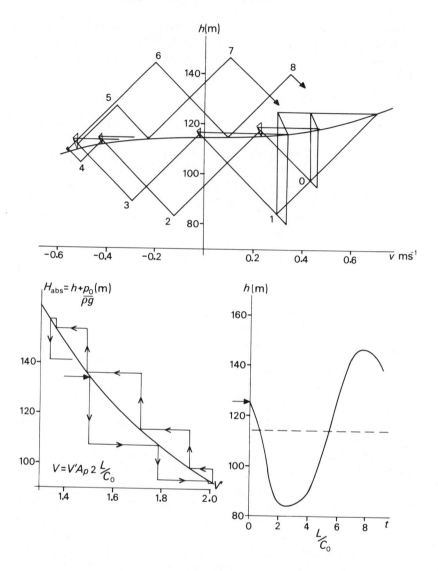

Fig. 13.1

prudent to omit an isolating valve between the pipeline and the air vessel. Finally, very high velocities may occur in the connecting pipe and some thought should be given to the internal surfaces of adjacent parts subjected to these velocities.

13.2.4   *What is the effect of a throttle connection to an air vessel?*

It has been found that economies, e.g. reduction in the size of tank, can be achieved by providing some form of constriction between the surge tank (open or closed) and the pipeline. Most methods have arisen from an empirical approach which has involved comprehensive field tests, and a far sighted authority will always make provision for these.

In the case of a pumping plant installation the insertion of an orifice, for the same size tank, results in approximately the same downsurge on pump failure – discharge side, but less ultimate overpressure. On the other hand, if

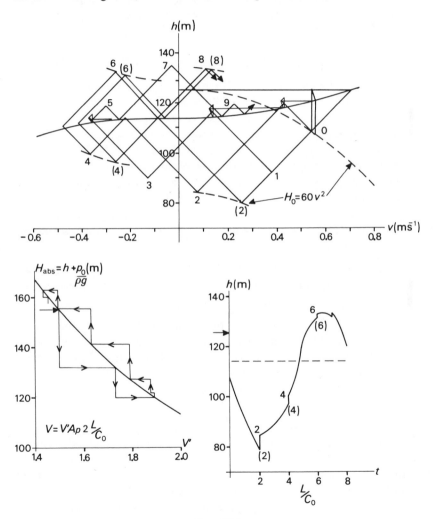

Fig. 13.2

an air vessel is reduced in size because the throttle requires less air volume for the same overpressure it must be realized that greater downsurge in the pipe is likely. This aspect is becoming of greater importance since downsurge is generally the reason for water column rupture. One must reconcile the gains of reduced air vessel size against increased risk of column rupture. As an example of a throttle with an air vessel, the results of an analysis of an extension of the previous example are illustrated in Fig. 13.2. The losses through an orifice with very high velocities transiently imposed have been determined empirically and a loss of 60 v^2 was allowed. It should be noted that some designers also incorporate an orifice which is intended to have different losses for forward and reverse flow.

13.2.5   Show how bypass methods may be used to reduce water hammer.

Since the development of excessive water hammer pressures is due to sudden changes of the velocity of flow, protection may be achieved by slowly decelerating the liquid. Thus, for example, when flow reverses towards a pump a bypass may be opened allowing water to flow to the suction side. The bypass valve then may be closed slowly with only a small overpressure developing. Similarly, in a hydroelectric system, when regulation requires a sudden decrease in output a bypass may open rapidly in concert with the flow reduction through the turbine.

In the same category are pressure relief valves which open very rapidly as warning of an excessive pressure increase occurs. It is clear that two factors must be considered:

(1)   The valve must be capable of rapid control with known characteristics.
(2)   The discharge through the bypass constitutes a waste of energy, and the disposal of the liquid must be provided for.

An example of the operation of a bypass is now illustrated graphically in Fig. 13.3. It should be noted that the various valve positions are labelled with T values, and the importance of relating these to actual valve control positions should not be underestimated.

13.2.6   Are there other methods for protection?

This introductory text will not treat in detail all those methods ranging from elaborate surge suppressors to the tuning of a systems frequencies with a short branch with dead end, a device common in household plumbing. However, there are several types that should be mentioned without further treatment to indicate the diversity of the methods that are available.

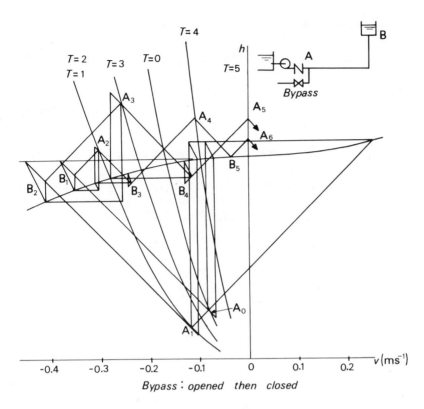

Fig. 13.3

(1) One way surge tank   When pressures less than atmospheric occur in a pipe, a surface reservoir isolated by a non-return valve can automatically discharge into the pipeline. This is the reverse of the bypass method of Problem 13.2.5. It would be suitable for long pipelines since time must be allowed for sufficient volume to flow. Methods for refilling must be considered (see Miyashiro, 1967; Stephenson, 1972).

(2) Controlled closing of a non-return valve   By special servocontrol or a damping mechanism the non-return valve closure may be controlled. This draws attention to the very important consideration of the continued satisfactory operation of all valve mechanisms associated with the system. Very few problems are as serious as the delayed closing of a non-return valve due to spindle friction. In a large pumping plant an automatic (uncontrolled) device of this kind might not be tolerated and a hydraulic power operated isolating valve used instead.

(3) **Antivacuum valves** Air may be admitted when pressures are less than atmospheric and prevent low pressures from occurring. This form of control has also been considered by many as a way of reducing the severity of pressures when the water column rejoins replacing the vapour cavity with an air cushion. It is necessary to weigh the advantages of this form of low pressure control against the disadvantages of air accumulation and subsequent removal.

(4) **Valve stroking** A deliberate phasing of the degree of opening of a valve so as to control the ultimate water hammer pressures is referred to as valve stroking. It assumes a detailed knowledge of a valve characteristic, particularly when nearly closed, and requires servomechanisms for operation (see Streeter and Wylie, 1967; Kinno, 1967).

14

Resonance – impedance method

14.1 Introduction

Whereas water hammer has been considered thus far as a shock phenomenon the disturbances observed in pipe systems may derive from oscillations in the flow which rapidly build up to destructive magnitudes. Thus water hammer may be legitimately confused with a resonance effect due to the amplification of the oscillations. It is often essential, therefore, that the features of the system be studied to see if such effects are possible. Resonance in hydropower systems has been reported (Jaeger, 1965) and can be a serious impediment to satisfactory regulation and control.

It is well known that the mathematical solution of systems subjected to oscillation may be accomplished by the complex representation of periodic phenomena. The impedance method has been well-illustrated by E. J. Waller in a series of publications from Oklahoma State University (1958, 1959) and more recently E. B. Wylie (1965) has demonstrated the method in complex pipe networks (see also Zielke and Hack, 1972).

The formal solution procedure involves the application of a sinusoidal oscillation, the frequency of which is varied over a useful range and the amplitude and phase characteristics of all parts of the system are determined. The system behaviour is expressible in the form

$$h = zq$$

where z is the complex impedance connecting the oscillating pressure head to an oscillating flow q. The value of z is determined for each frequency and when it is large a small oscillating q clearly produces a large oscillating h.

The basic water hammer equations, (2.4) and (2.19), written here for convenience with Q as a principal variable rather than v, are:

$$g\frac{\partial h}{\partial x} + \frac{1}{A}\frac{\partial Q}{\partial t} + \frac{\tau_0}{\rho R} = 0$$

$$\frac{C_0^2}{gA}\frac{\partial Q}{\partial x} + \frac{\partial h}{\partial t} = 0$$

The pipeline is assumed horizontal and the usual approximation that

$v\partial h/\partial x$, $v\partial v/\partial x$ are small compared with $\partial h/\partial t$ and $\partial y/\partial t$, respectively, is made.

The theory is concerned with small oscillations and thus in addition to the simplification that the area is invariant the substitutions for h and Q are made as follows, with bar denoting mean quantity:

$$h = \overline{h} + h'; \quad Q = \overline{Q} + Q' \tag{14.1}$$

Steady and unsteady friction are assumed to follow the quadratic law and, after substitution and cancellation of the derivatives of mean quantities, the water hammer equations are then modified to read:

$$\frac{\partial h'}{\partial x} + \frac{1}{gA}\frac{\partial Q'}{\partial t} + rQ' = 0 \tag{14.2}$$

$$\frac{\partial Q'}{\partial x} + \frac{gA}{C_0^2}\frac{\partial h'}{\partial t} = 0 \tag{14.3}$$

where $r = f\overline{Q}/gdA^2$ and f is the friction coefficient from the Darcy–Weisbach formula. Differentiation of the modified equations gives:

$$\frac{\partial^2 h'}{\partial x^2} = \frac{1}{C_0^2}\frac{\partial^2 h'}{\partial t^2} + \frac{gAr}{C_0^2}\frac{\partial h'}{\partial t} \tag{14.4}$$

A similar equation may be obtained with Q' as a function of x and t.

The separation of variables method is used to integrate the second order partial differential equation and as we are concerned with periodic fluctuations the complex impedance solution is sought. If it is assumed that $h' = X(x)T(t)$ where $X(x)$ and $T(t)$ are functions respectively of x and t only, Equation (14.4) becomes

$$\frac{1}{X}\frac{d^2 X}{dt^2} = \frac{1}{T}\left(\frac{1}{C_0^2}\frac{d^2 T}{dt^2} + \frac{gAr}{C_0^2}\frac{dT}{dt}\right) \tag{14.5}$$

Since this can only be true if each side of the equation is equal to the same constant, then, introducing $\gamma = \alpha + i\beta$,

$$\frac{1}{X}\frac{d^2 X}{dt^2} = \gamma^2 \tag{14.6}$$

The solution of Equation (14.6) is

$$X = A_1 e^{\gamma x} + B_1 e^{-\gamma x} \tag{14.7}$$

where A_1 and B_1 are constants.

In the second equality from Equation (14.5) and introducing $T = Ce^{i\omega t}$

$$\gamma^2 = \frac{Ag\omega}{C_0^2}\left(-\frac{\omega}{gA} + ir\right) \tag{14.8}$$

which expresses γ, known as the propagation constant, in terms of the parameters of the system for an harmonic type of oscillation. The value of h' becomes:

$$h' = e^{i\omega t} (C_1 e^{\gamma x} + C_2 e^{-\gamma x}) \tag{14.9}$$

where $C_1 = A_1 C$ and $C_2 = B_1 C$. Substituting Equation (14.9) in Equations (14.2) and (14.3),

$$Q' = \frac{\omega g A}{i C_0^2 \gamma} e^{i\omega t} (C_1 e^{\gamma x} + C_2 e^{-\gamma x}) \tag{14.10}$$

The fundamental equation relating h' and Q' is required in the form

$$h' = ZQ' \tag{14.11}$$

in which the complex impedance Z is

$$Z = \frac{i\gamma C_0^2}{\omega g A} \left(\frac{C_1 e^{\gamma x} + C_2 e^{-\gamma x}}{C_1 e^{\gamma x} - C_2 e^{-\gamma x}} \right) \tag{14.12}$$

The formal arrangement of the water hammer equations into a solution for small oscillatory motions superimposed upon a steady state flow has now been achieved in terms of a complex impedance Z. Some of the properties of Z and the evaluation of boundary conditions will be outlined prior to some elementary examples.

14.2   Problems

14.2.1   What is the relevance of the propagation constant?

The complex impedance is a function of the circuit constants, the frequency of oscillation and the propagation. The propagation constant γ is complex and the real numbers α and β can be derived as follows:

$$\gamma^2 = (\alpha + i\beta)^2 = (|z| (\cos \phi + i \sin \phi))^2$$
$$= |z|^2 (\cos 2\phi + i \sin 2\phi) = W \tag{14.13}$$

W (see Fig. 14.1) will be in the second quadrant because it has a negative real part and a positive imaginary part (see Equation (14.8)), and $\pi/2 < 2\phi < \pi$ and hence $0 < \phi < \pi/2$.

Equating real and imaginary parts in Equations (14.8) and (14.13) and simplifying,

$$\tan 2\phi = -\frac{Agr}{\omega} = -\tan (\pi - 2\phi)$$

and thus

$$\phi = \frac{\pi}{2} - \frac{1}{2}\tan^{-1}\frac{Agr}{\omega} \tag{14.14}$$

It is readily established that

$$\alpha = \frac{\omega}{C_0}\left(1 + \left(\frac{Agr}{\omega}\right)^2\right)^{\frac{1}{4}}\sin\left(\frac{1}{2}\tan^{-1}\frac{Agr}{\omega}\right) \tag{14.15}$$

and

$$\beta = \frac{\omega}{C_0}\left(1 + \left(\frac{Agr}{\omega}\right)^2\right)^{\frac{1}{4}}\cos\left(\frac{1}{2}\tan^{-1}\frac{Agr}{\omega}\right) \tag{14.16}$$

Alternative expressions for α and β are:

$$\alpha = \frac{\omega}{\sqrt{2C_0}}\left(\sqrt{1 + \left(\frac{Agr}{\omega}\right)^2} - 1\right)^{\frac{1}{2}} \tag{14.15a}$$

$$\beta = \frac{\omega}{\sqrt{2C_0}}\left(\sqrt{1 + \left(\frac{Agr}{\omega}\right)^2} + 1\right)^{\frac{1}{2}} \tag{14.16a}$$

In the case of a frictionless system $r = 0$ and thus $\alpha = 0$ and $\beta = \omega/C_0$ and the propagation constant becomes

$$\gamma = i\omega/C_0 \tag{14.17}$$

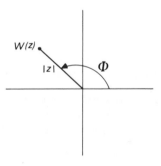

Fig. 14.1

14.2.2   What particular form does the complex impedance take?

In the application to electrical circuits the designer attempts a process of impedance matching, which is tantamount to avoiding undesirable reflections from terminal parts of the circuit. The result is equivalent in transmission lines to an infinite length of line. In the hydraulic application and with an infinite line the complex impedance Z becomes, by definition, the characteristic impedance:

$$Z_c = \frac{i\gamma C_0{}^2}{\omega g A} \qquad (14.18)$$

In the $+x$ direction $Z = -Z_c$, and the $-x$ direction $Z = +Z_c$.

The complex (hydraulic) impedance at any location x may be written,

$$Z = -Z_c \frac{(C_1 e^{\gamma x} + C_2 e^{-\gamma x})}{(C_1 e^{\gamma x} - C_2 e^{-\gamma x})} \qquad (14.19)$$

The values of C_1 and C_2 (integration constants) will depend on the boundary conditions. Terminal condition $x = 0$ yields (see Equations (14.9) and (14.10)):

$$h_R{}' = e^{i\omega t}(C_1 + C_2) \quad \text{and} \quad -Z_c Q_R{}' = e^{i\omega t}(C_1 - C_2)$$

Combining these:

$$C_1 = \tfrac{1}{2}e^{-i\omega t}(h_R{}' - Z_c Q_R{}') \qquad (14.20)$$

and

$$C_2 = \tfrac{1}{2}e^{-i\omega}(h_R{}' + Z_c Q_R{}') \qquad (14.21)$$

With the above values of C_1 and C_2 substituted in Equation (14.19) and reducing to hyperbolic functions the complex impedance at position x becomes:

$$Z = \frac{Z_R - Z_c \tanh \gamma x}{1 - \dfrac{Z_R}{Z_c} \tanh \gamma x} \qquad (14.22)$$

When $x = L$, $Z = Z_S$, the subscript S designating sending end, and when $x = 0$, $Z = Z_R$ and R refers to the receiving end.

It can be readily shown by rearrangement that

$$Z_R = \frac{Z_S + Z_c \tanh \gamma L}{1 + \dfrac{Z_S}{Z_c} \tanh \gamma L} \qquad (14.23)$$

14.2.3 Describe the boundary condition forms of the complex impedance.

(1) Constant head reservoir at $x = 0$. Here $h' = 0$ and thus:

$$Z_R = 0 \qquad (14.24)$$

$$Z(x) = -Z_c \tanh \gamma x \qquad (14.25)$$

In the case of a frictionless system $r = 0$, hence

$$Z(x) = -iZ_c \tan (\omega x / C_0) \qquad (14.26)$$

(2)   Junction of pipes in series (see Fig. 14.2):

$$h_{S_1} = h_{R_2} \quad \text{and} \quad Q_{S_1} = Q_{R_2}$$

Therefore

$$Z_{S_1} = Z_{R_2} \tag{14.27}$$

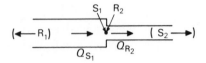

Fig. 14.2

(3)   Tee junction see Fig. 14.3):

$$h_{S_1} = h_{S_2} = h_{R_3} \text{ and continuity requires } Q_{S_1} = Q_{S_2} + Q_{R_3}$$

Therefore

$$\frac{1}{Z_{S_1}} = \frac{1}{Z_{S_2}} + \frac{1}{Z_{R_3}}$$

Rearranging the above the impedance then may be expressed as:

$$Z_{R_3} = \frac{Z_{S_2} Z_{S_1}}{Z_{S_2} - Z_{S_1}} \tag{14.28}$$

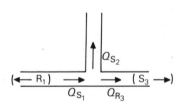

Fig. 14.3

(4)   A reservoir with a varying head has a finite Z since $h' \neq 0$. The boundary condition is derived as follows. In terms of the nth harmonic ω_n and corresponding phase angle ϕ, the fluctuating head may be written:

$$h_R(x) = \overline{h} e^{i(\omega_n t + \phi - \frac{\pi}{2})} \tag{14.29}$$

The free surface is a nodal point for pressure and the discharge is $180°$ out of phase there, but $Q = vA = \dfrac{dH}{dt} A$ and therefore

$$Q_R(x) = \overline{h}i\omega_n A\,e^{i(\omega_n t + \phi - \frac{\pi}{2})} = h_R(x)i\omega_n A$$

The end impedance is the ratio of the fluctuating head and discharge and thus:

$$Z_R = -\frac{i}{\omega_n A} \tag{14.30}$$

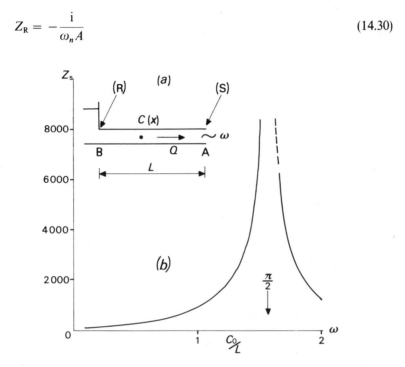

Fig. 14.4

14.2.4 Develop a computer solution for a single pipeline.

In Fig. 14.4a, an infinite reservoir is assumed upstream and the general nomenclature is illustrated. The impedance at a point C at x from B is according to Equation (14.22), which in the case of fixed reservoir level reduces to Equation (14.25).

In the PROGRAM IMP (Fig. P.10), the characteristic impedance (Z_c) is:

$$Z_c = \frac{C_0{}^2}{\omega g A}(\beta - i\alpha) = \frac{1}{\omega C_1}(\beta - i\alpha) \tag{14.31}$$

From Equation (14.25) the hydraulic impedance Z_S is:

$$Z_S = -(Z_{CR} + iZ_{CI})(T_{HR} + iT_{HI}) \tag{14.32}$$

```
                    PROGRAM IMP
          C         **  D=DIAMETER, F=FRICTION FACTOR   **
          C         **  G=GRAVITY, Q=DISCHARGE   **
                    AR = 0.7853982*D**2
                    R = F*Q/(G*D*AR*AR)
                    EL = 1.0/(G*AR)
                    C1 = G*AR/CO**2
          C         DETERMINATION OF CHARACTERISTIC IMPEDANCE
                    DO 10 I=2,20
                    OM = FLOAT(I)/10.0
                    X = R/(EL*OM)
                    EO1 = SQRT((EL*OM)**2+R**2)
                    EO = SQRT(EO1)
                    E1 = 0.5*ATAN(X)
                    E2 = SQRT(C1*OM)
                    E3 = SIN(E1)
                    AL = EO*E2*E3
                    E4 = COS(E1)
                    BE = EO*E2*E4
                    ZCR = BE/(C1*OM)
                    ZCI = -AL/(C1*OM)
                    Y = AL*FLOAT(LL)
                    Z = BE*FLOAT(LL)
                    F1 = SINH(Y)*COS(Z)
                    F2 = COSH(Y)*SIN(Z)
                    F3 = COSH(Y)*COS(Z)
                    F4 = SINH(Y)*SIN(Z)
                    THR = (F1*F3+F2*F4)/(F3**2+F4**2)
                    THI = (F2*F3-F1*F4)/(F3**2+F4**2)
                    ZSR(I) = -(ZCR*THR-ZCI*THI)
                    ZSI(I) = -(ZCI*THR+THI*ZCR)
                    ZS(I)=SQRT(ZSR(I)**2+ZSI(I)**2)
                    PHIQ(I) = ATAN2(ZSI(I),ZSR(I))
                    T(I) = 6.2831853/OM
                10  CONTINUE
                    CALL EXIT
                    END
```

Fig. P.10

where the letters R, I refer to the real and imaginary parts. In the program the modulus of Z_S is the square root of the sum of the squares of the real and imaginary parts. The phase angle, ϕ, can also be calculated from the ratio of the real and imaginary parts. AR is the pipe area and R is defined in Equation (14.2).

The result is shown graphically in Fig. 14.4b for a numerical problem with the following data: $C_0 = 980 \text{ m s}^{-1}$, $L = 980$ m, $D = 0.46728$ m, $f = 0.013$, $Q = 0.0283 \text{ m}^3 \text{ s}^{-1}$, ω varying from 0.2 to 2.0 rad s^{-1}. In this case it is clear that the resonant period should be $4L/C_0 = 4$ secs and the example confirms that oscillations of discharge at A (the gate end) at $\pi/2$ rad s^{-1} would produce very large pressure head fluctuations.

```
PROGRAM IMP2
C    **   AR1, AR2 =AREAS. Q1, Q2 =DISCHARGES   **
     DIMENSION ZS1(300),ZS2(300),PHIQ1(300),PHIQ2(300),T(300)
     DIMENSION ZC1(300),ZC2(300),RZCS1(300),RZCS2(300)
     DIMENSION SR1(300),SI1(300),SR2(300),SI2(300)
     D1 = SQRT(AR1/0.7853982)
     D2 = SQRT(AR2/0.7853982)
     R1 = F1*Q1/(G*D1*AR1**2)
     R2 = F2*Q2/(G*D2*AR2**2)
     EL1 = 1.0/(G*AR1)
     EL2 = 1.0/(G*AR2)
     C11 = G*AR1/(CO1**2)
     C12 = G*AR2/(CO2**2)
     DO 7 I=2,MAXO,2
     OM = FLOAT(I)/10.0
     CALL ZCH(OM,R1,EL1,C11,AL1,BE1,ZCR1,ZCI1)
     CALL TRIG(AL1,BE1,LL1,F11,F21,F31,F41)
     CALL THRI(F11,F21,F31,F41,THR1,THI1)
     CALL ZIMP(-ZCR1,-ZCI1,THR1,THI1,ZSR1,ZSI1)
     SR1(I)=ZSR1
     SI1(I)=ZSI1
     CALL ZCH(OM,R2,EL2,C12,AL2,BE2,ZCR2,ZCI2)
     CALL TRIG(AL2,BE2,LL2,F12,F22,F32,F42)
     CALL THRI(F12,F22,F32,F42,THR2,THI2)
     CALL ZIMP(ZCR2,ZCI2,THR2,THI2,C1,C2)
     CALL ZIMP(ZSR1,ZSI1,THR2,THI2,C3,C4)
     CALL THRI(ZSR1-C1,ZSI1-C2,ZCR2-C3,ZCI2-C4,C5,C6)
     CALL ZIMP(ZCR2,ZCI2,C5,C6,ZSR2,ZSI2)
     SR2(I)=ZSR2
     SI2(I)=ZSI2
     ZS1(I) = SQRT(ZSR1**2+ZSI1**2)
     ZS2(I) = SQRT(ZSR2**2+ZSI2**2)
     PHIQ1(I) = 0.0
     PHIQ2(I) = 0.0
     T(I) = 6.2831853/OM
     ZC1(I) = SQRT(ZCR1**2+ZCI1**2)
     ZC2(I) = SQRT(ZCR2**2+ZCI2**2)
     RZCS1(I) = ZS1(I)/ZC1(I)
7    RZCS2(I) = ZS2(I)/ZC2(I)
     CALL EXIT
     END
```

Fig. P.11a

14.2.5   Develop the solution for two pipes in series.

The condition at the junction of two pipes is indicated in Equation (14.27) and this enables the specified boundary conditions (terminal impedance) to be used for solving the whole system. At the downstream end the hydraulic impedance becomes:

$$Z_{S_2} = \frac{Z_{S_1} Z_{C_2} - \left(Z_{C_2}\right)^2 \tanh \gamma L_2}{Z_{C_2} - Z_{S_1} \tanh \gamma L_2} \tag{14.33}$$

The solution of these more complicated equations is conveniently achieved with subroutines.

```
SUBROUTINE THRI(X1,Y1,X2,Y2,X3,Y3)
X3 = (X1*X2+Y1*Y2)/(X2**2+Y2**2)
Y3 = (Y1*X2-X1*Y2)/(X2**2+Y2**2)
RETURN
END




SUBROUTINE ZIMP(X1,Y1,X2,Y2,X3,Y3)
X3 = X1*X2-Y1*Y2
Y3 = Y1*X2+Y2*X1
RETURN
END




SUBROUTINE ZCH(OM,R,EL,C1,AL,BE,ZCR,ZCI)
X = R/(EL*OM)
EO1 = SQRT((EL*OM)**2+R**2)
EO = SQRT(EO1)
E1= 0.5*ATAN(X)
E2 = SQRT(C1*OM)
E3 = SIN(E1)
E4 = COS(E1)
AL = EO*E2*E3
BE = EO*E2*E4
ZCR = BE/(C1*OM)
ZCI = -AL/(C1*OM)
RETURN
END




SUBROUTINE TRIG(AL,BE,LL,F1,F2,F3,F4)
Y = AL*FLOAT(LL)
Z = BE*FLOAT(LL)
F1 = SINH(Y)*COS(Z)
F2 = COSH(Y)*SIN(Z)
F3 = COSH(Y)*COS(Z)
F4 = SINH(Y)*SIN(Z)
RETURN
END
```

Fig. P.11b

In the PROGRAM IMP2 (Fig. P.11), the solution for the no friction case is obtained by putting PHIQ = 0. The range of frequencies can be extended as far as desired by the choice of MAXO, the maximum value of omega. A reservoir is placed upstream and the program omits the formal input and output statements.

A numerical example is worked (ignoring friction) for two pipes where $L_1 = L_2 = 305$ m. One result shown in Fig. 14.5 is the variation of the hydraulic impedance at the downstream end for a range of ω from 0.2 to

20 rad s^{-1}. The remaining data is: $C_0 = 1280$ m s^{-1}, $D_1 = 0.40988$ m, $D_2 = 0.34393$, and $Q = 0.0283$ m^3 s^{-1}. As in the previous example one would expect the lowest resonant frequency to be $\Sigma 4L/C_0$ which is, in this example, 1.906 s. In fact it is nearer 1.745 s.

This result indicates one of the justifications for detailed calculation of the resonant characteristics of a pipe system. Even in this simple illustration it is not possible to predict (by simple means) the fundamental resonant frequency of a complex system (see later).

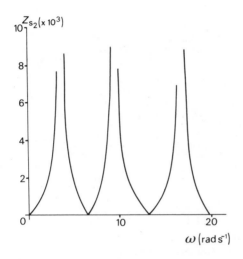

Fig. 14.5

14.2.6   Demonstrate the basic program steps for a system incorporating a branch.

At a tee, the hydraulic impedance of one pipe is expressed in terms of the other as shown in Equation (14.28). In PROGRAM IMP3 (Fig. P.12) the case is considered of an upstream reservoir (water level constant) on pipe 1 and the branch as pipe 2 is acting as a standpipe with an area unchanged from the branch area in which allowance is made for the variation of the water level according to Equation (14.30).

The program, Fig. P.12, uses the same SUBROUTINES as those in PROGRAM IMP2 (Fig. P.11b). Friction is again zero by putting PHIQ = 0.

```
      PROGRAM IMP3
C     AR1, AR2, AR3 =AREAS. C01, C02, C03 =WAVE SPEEDS **
C     ** Q1, Q2, Q3 =DISCHARGES **
      D1 = SQRT(AR1/0.7853982)
      D2 = SQRT(AR2/0.7853982)
      D3 = SQRT(AR3/0.7853982)
      R1 = F1*Q1/(G*D1*AR1**2)
      R2 = F2*Q2/(G*D2*AR2**2)
      R3 = F3*Q3/(G*D3*AR3**2)
      EL1 = 1.0/(G*AR1)
      EL2 = 1.0/(G*AR2)
      EL3 = 1.0/(G*AR3)
      C11 = G*AR1/(C01**2)
      C12 = G*AR2/(C02**2)
      C13 = G*AR3/(C03**2)
      DO 7 I=2,MAXO,2
      OM = FLOAT(I)/10.0
      CALL ZCH(OM,R1,EL1,C11,AL1,BE1,ZCR1,ZCI1)
      CALL TRIG(AL1,BE1,LL1,F11,F21,F31,F41)
      CALL THRI(F11,F21,F31,F41,THR1,THI1)
      CALL ZIMP(-ZCR1,-ZCI1,THR1,THI1,ZSR1,ZSI1)
      SR1(I)=ZSR1
      SI1(I)=ZSI1
      CALL ZCH(OM,R2,EL2,C12,AL2,BE2,ZCR2,ZCI2)
      CALL TRIG(AL2,BE2,LL2,F12,F22,F32,F42)
      CALL THRI(F12,F22,F32,F42,THR2,THI2)
      ZRR2=0.0
      ZRI2=-1.0/(OM*AR2)
      CALL ZIMP(ZCR2,ZCI2,THR2,THI2,B1,B2)
      CALL ZIMP(ZRR2,ZRI2,THR2,THI2,B3,B4)
      CALL THRI(ZRR2-B1,ZRI2-B2,ZCR2-B3,ZCI2-B4,B5,B6)
      CALL ZIMP(ZCR2,ZCI2,B5,B6,ZSR2,ZSI2)
      SR2(I)=ZSR2
      SI2(I)=ZSI2
      CALL ZIMP(ZSR2,ZSI2,ZSR1,ZSI1,C1,C2)
      CALL THRI(C1,C2,ZSR2-ZSR1,ZSI2-ZSI1,ZRR3,ZRI3)
      RR3(I) = ZRR3
      RI3(I) = ZRI3
      CALL ZCH(OM,R3,EL3,C13,AL3,BE3,ZCR3,ZCI3)
      CALL TRIG(AL3,BE3,LL3,F13,F23,F33,F43)
      CALL THRI(F13,F23,F33,F43,THR3,THI3)
      CALL ZIMP(ZCR3,ZCI3,THR3,THI3,E1,E2)
      CALL ZIMP(ZRR3,ZRI3,THR3,THI3,E3,E4)
      CALL THRI(ZRR3-E1,ZRI3-E2,ZCR3-E3,ZCI3-E4,E5,E6)
      CALL ZIMP(ZCR3,ZCI3,E5,E6,ZSR3,ZSI3)
      SR3(I)=ZSR3
      SI3(I)=ZSI3
      ZS1(I) = SQRT(ZSR1**2+ZSI1**2)
      ZS2(I) = SQRT(ZSR2**2+ZSI2**2)
      ZS3(I)=SQRT(ZSR3**2+ZSI3**2)
      PHIQ1(I) = 0.0
      PHIQ2(I) = 0.0
      PHIQ3(I) = 0.0
      T(I) = 6.2831853/OM
      ZC1(I) = SQRT(ZCR1**2+ZCI1**2)
      ZC2(I) = SQRT(ZCR2**2+ZCI2**2)
      ZC3(I) = SQRT(ZCR3**2+ZCI3**2)
      RZCS1(I) = ZS1(I)/ZC1(I)
      RZCS2(I) = ZS2(I)/ZC2(I)
      RZCS3(I) = ZS3(I)/ZC3(I)
7     CONTINUE
      CALL EXIT
      END                    Fig. P.12
```

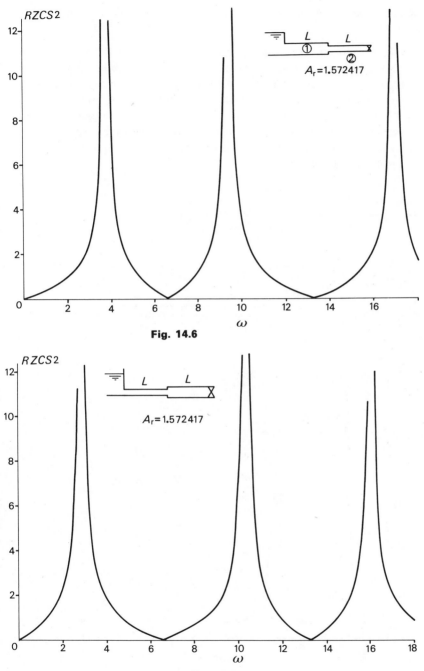

Fig. 14.6

Fig. 14.7

14.2.7   What is the fundamental resonant frequency for several pipes in series?

Consider two pipes in series of equal length (say 305 m) with an area ratio of 1.572417. With an upstream reservoir there are two possible configurations, diverging or converging. The immediate solution for either case would seem to be a fundamental frequency corresponding to $\Sigma \, 4L/C_0$ which is 0.785 C_0/L (see Streeter and Wylie, 1967). The analysis of both cases has been performed, ignoring friction for convenience and the results are illustrated in Figs 14.6 and 14.7. In Table 14.1 the first three critical frequencies for each system have been listed.

Table 14.1   Critical frequencies for two systems of pipes

	Critical frequency (C_0/L units)		
	1st	2nd	3rd
Converging pipes	0.898	2.24	4.03
Diverging pipes	0.673	2.47	3.82

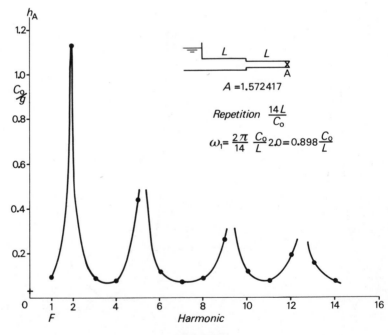

Fig. 14.8

In both cases the lowest resonant frequency differs markedly from $0.785\, C_0/L$. In general it is not possible to predict the resonant frequencies of circuits without a detailed numerical analysis. However, in the simplest case of two pipes of equal length a thorough study of the dynamic behaviour has enabled some empirical results to be derived.

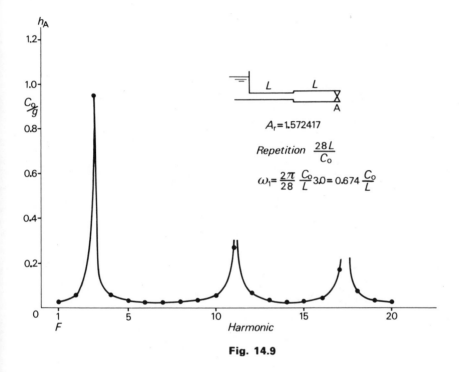

Fig. 14.9

Table 14.2 Critical frequency values

	Critical ω (C_0/L units)		
Solution	1st	2nd	3rd
Impedance (Fig. 14.8)	0.898	2.24	4.03
Converging Fourier	0.898	2.30	4.08
Harmonic [•]	(4)	(10)	(18)
Impedance (Fig. 14.9.)	0.673	2.47	3.83
Diverging Fourier	0.674	2.48	3.82
Harmonic [•]	(3)	(11)	(17)

[•] Harmonic = approximate harmonic derived from the spectrum of frequencies

In the examples just considered, the results were for a cyclic disturbance. If the disturbance was a spectrum of frequencies determined by decomposing a step function, then the two-pipe system can be shown to be members of the same family and the critical frequencies no longer seem unrelated. In Figs 14.8 and 14.9, the various critical frequencies for the step function response can be derived and when listed in Table 14.2 the critical frequencies may be regarded as 'harmonics' of a 'fundamental' $0.224L/C_0$.

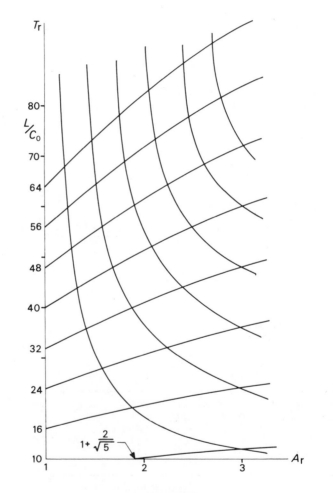

Fig. 14.10

Some results of analysis to determine the common fundamental for two pipes of equal length, diverging or converging, have been incorporated in graphical form in Fig. 14.10. The fundamental period T_r for a number of area ratios A_r can be found from the intersection of the two families of curves. For example, with $T_r = 64L/C_0$, area ratios of 1.511 and 2.240 would present a series of critical frequencies that would appear as harmonics of that fundamental period whether the pipes are diverging or converging in the downstream direction of the base flow.

Appendix 1

Wave speed

In Chapter 2, the basic equations for unsteady flow were developed in simplified form and the main features indicated. A significant parameter in the motion is the wave speed or celerity, C_0, expressed by Equation (2.15). Note that the theory has been restricted to the case of the linear approximation for wave speed.

The way that C_0 is affected by the physical restraints of the pipeline can be evaluated in certain extreme cases. In liquids the stress due to pressure (p) is the same in all directions whereas in solids a stress (or strain) which may be introduced in a specific direction induces a stress (or strain) in other directions, depending on the bulk properties of the solid.

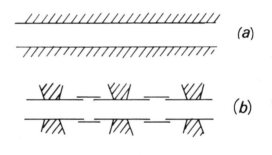

(a)

(b)

Fig. A1.1

An idealization of the pipeline is shown in Fig. A1.1. In case (a) the pipeline is anchored throughout its entire length and in case (b) it is anchored in between expansion joints. It will be evident later that Equation (2.15) corresponds to case (b). The evaluation of these cases is dependent also on the ratio of pipe thickness to pipe diameter.

In Fig. A1.2 the primary stress is due to hoop tension and if a stress variation is considered in the wall, i.e. the wall thickness, e, is significant in relation to D, then the stresses are:

$$\sigma_r = p \frac{D_1^{\,2}}{D_2^{\,2} - D_1^{\,2}} \left(1 - \frac{D_2^{\,2}}{D^2}\right) \tag{A1.1}$$

and circumferentially

$$\sigma_c = p \frac{D_1{}^2}{D_2{}^2 - D_1{}^2}\left(1 + \frac{D_2{}^2}{D^2}\right) \tag{A1.2}$$

The longitudinal stress will be

$$\sigma_1 - \mu(\sigma_r + \sigma_c) = \sigma_1 - 2p\mu\left(\frac{2D_1{}^2}{D_2{}^2 - D_1{}^2}\right) \tag{A1.3}$$

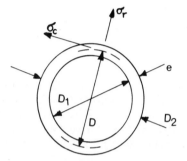

Fig. A1.2

which consists of two contributions, that due to longitudinal restraint and that due to transverse stress. When the longitudinal strain is constant the longitudinal stress does not vary across the pipe wall.

The strains in the three directions then become:

$$\varepsilon_r = \frac{p}{E}\frac{D_1{}^2}{D_2{}^2 - D_1{}^2}\left((1 - \mu) - \frac{D_2{}^2}{D^2}(1 + \mu)\right) - \frac{\mu\sigma_1}{E} \tag{A1.4}$$

$$\varepsilon_c = \frac{p}{E}\frac{D_1{}^2}{D_2{}^2 - D_1{}^2}\left((1 - \mu) + \frac{D_2{}^2}{D^2}(1 + \mu)\right) - \frac{\mu\sigma_1}{E} \tag{A1.5}$$

$$\varepsilon_1 = \frac{\sigma_1}{E} - 2p\mu\frac{D_1{}^2}{D_2{}^2 - D_1{}^2} \tag{A1.6}$$

For thin wall theory, the conditions at the inner wall are first found $(D = D_1)$, and then as D_2 approaches D_1 the hoop stress becomes:

$$\frac{D_1 + e}{2e}p + \mu p - \sigma_1 \tag{A1.7}$$

The simplified theory considers neither longitudinal stress nor the Poisson effect and thus Equation (A1.7) reduces to the first term only and usually e is neglected in relation to D_1 for a thin wall.

The cases considered give the hoop stress as follows:

(a) Longitudinal strain is zero

$$\left(\frac{D_1 + e}{2e}(1 - \mu^2) + \mu(1 + \mu) \right)p$$

approximated by $\dfrac{D_1}{2e}(1 - \mu^2)$.

(b) Longitudinal stress is zero

$$\left(\frac{D_1 + e}{2e} + \mu \right)p$$

approximated by $\dfrac{D_1}{2e}$.

Practical values

The values of K and ρ for water vary with temperature and the variation is shown in Fig. A1.3. Note that K is also a function of absolute pressure (see *The Mechanical Properties of Fluids* by Drysdale, Gibson and Taylor, 1944). Approximate values of E for different materials in Table A1.1 enable an estimate of C_0 to be determined for a liquid, given the pipeline geometry. In Fig. A1.4 the variation of C_0 is depicted for selected values of E with D/e as a basic variable, for the case (b) (above) where longitudinal stress is zero.

Table A1.1 Approximate values of E (Young's modulus)

Material	$E\ (\times 10^{-9})$
Aluminium alloy	70
Asbestos cement	24
Brass	100
Concrete	20
Copper	120
Glass	70
Iron	100
Lead	10
Plastic●	
Polyethylene	0.8
Polystyrene	5
PVC	2.7
Steel	210
Rock	
Granite	50
Sandstone	3

● Value depends significantly on temperature

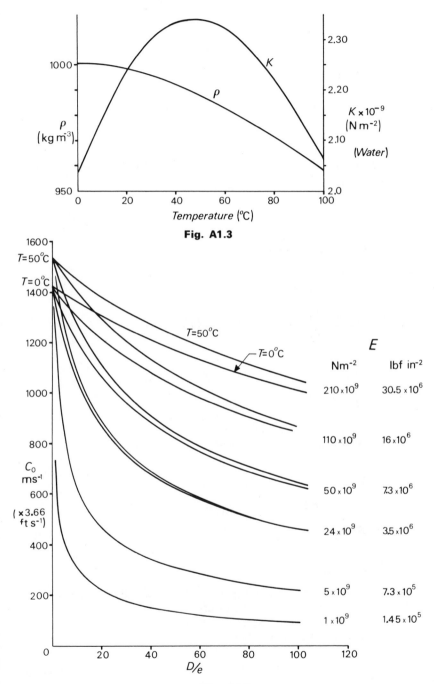

Fig. A1.3

Fig. A1.4

Appendix 2

The integral form of the basic equations – method of d'Alembert

The basic equation of wave motion is

$$\frac{\partial^2 \phi}{\partial t^2} = C_0{}^2 \frac{\partial^2 \phi}{\partial x^2} \tag{A2.1}$$

Changing to variables $u = x - C_0 t$, $v = x + C_0 t$, $\phi = f(u, v)$, then

$$\frac{\partial \phi}{\partial x} = \frac{\partial f}{\partial u}\frac{\partial u}{\partial x} + \frac{\partial f}{\partial v}\frac{\partial v}{\partial x} = \frac{\partial \phi}{\partial u} + \frac{\partial \phi}{\partial v}$$

$$\frac{\partial \phi}{\partial t} = \frac{\partial f}{\partial u}\frac{\partial u}{\partial t} + \frac{\partial f}{\partial v}\frac{\partial v}{\partial t} = -C_0 \left\{ \frac{\partial \phi}{\partial u} + \frac{\partial \phi}{\partial v} \right\}$$

$$\frac{\partial^2 \phi}{\partial x^2} = \frac{\partial^2 \phi}{\partial u^2} + \frac{2\partial^2 \phi}{\partial u\, \partial v} + \frac{\partial^2 \phi}{\partial v^2}$$

$$\frac{\partial^2 \phi}{\partial t^2} = C_0{}^2 \left\{ \frac{\partial^2 \phi}{\partial u^2} - \frac{2\partial^2 \phi}{\partial u\, \partial v} + \frac{\partial^2 \phi}{\partial v^2} \right\}$$

Substitution in Equation (A2.1) gives:

$$\frac{\partial^2 \phi}{\partial u\, \partial v} = 0 \tag{A2.2}$$

The most general solution of this equation is

$$\phi = f(u) + g(v)$$

that is:

$$\phi = f(x - C_0 t) + g(x + C_0 t) \tag{A2.3}$$

Thus an equation of the form of Equation (A2.3) reduces to the same result as Equations (2.20) and (2.21) when differentiated.

Now ϕ may be h or v which are related so different forms will be required. Therefore, it may be assumed that

$$\phi_1 = f_1 \left(t - \frac{x}{C_0} \right) + g_1 \left(t + \frac{x}{C_0} \right) \tag{A2.4}$$

$$\phi_2 = f_2\left(t - \frac{x}{C_0}\right) - g_2\left(t + \frac{x}{C_0}\right) \tag{A2.5}$$

where the arguments have been expressed differently as a time base. If these are differentiated as before:

$$\frac{\partial \phi_1}{\partial x} = -\frac{1}{C_0}\frac{\partial \phi_1}{\partial u} + \frac{1}{C_0}\frac{\partial \phi_1}{\partial v}$$

$$\frac{\partial \phi_2}{\partial x} = -\frac{1}{C_0}\frac{\partial \phi_2}{\partial u} - \frac{1}{C_0}\frac{\partial \phi_2}{\partial v}$$

$$\frac{\partial \phi_1}{\partial t} = \frac{\partial \phi_1}{\partial u} + \frac{\partial \phi_1}{\partial v}$$

$$\frac{\partial \phi_2}{\partial t} = \frac{\partial \phi_2}{\partial u} - \frac{\partial \phi_2}{\partial v}$$

and applying these to Equations (2.18) and (2.19), i.e.

$$g\frac{\partial h}{\partial x} + \frac{\partial v}{\partial t} = 0 \tag{Equation 2.18}$$

$$\frac{C_0{}^2}{g}\frac{\partial v}{\partial x} + \frac{\partial h}{\partial t} = 0 \tag{Equation 2.19}$$

where we use ϕ_1 for h and ϕ_2 for v, these become:

$$-\frac{g}{C_0}\left(\frac{\partial \phi_1}{\partial u} - \frac{\partial \phi_1}{\partial v}\right) - \frac{\partial \phi_2}{\partial u} + \frac{\partial \phi_2}{\partial v} = 0$$

$$-\frac{C_0}{g}\left(\frac{\partial \phi_2}{\partial u} + \frac{\partial \phi_2}{\partial v}\right) - \frac{\partial \phi_1}{\partial u} - \frac{\partial \phi_2}{\partial v} = 0$$

Combining and integrating:

$$\phi_1 = -\frac{C_0}{g}\phi_2 \tag{A2.6}$$

Thus a suitable general solution (ϕ_1 and ϕ_2), is:

$$h - h_0 = F\left(t - \frac{x}{C_0}\right) + f\left(t + \frac{x}{C_0}\right) \tag{Equation 2.24}$$

$$v - v_0 = -\frac{C_0}{g}\left[F\left(t - \frac{x}{C_0}\right) - f\left(t + \frac{x}{C_0}\right)\right] \tag{Equation 2.25}$$

Summation and addition of these equations yields:

$$h - h_0 = \frac{C_0}{g}(v - v_0) + 2F\left(t - \frac{x}{C_0}\right) \tag{Equation 2.26}$$

$$h - h_0 = -\frac{C_0}{g}(v - v_0) + 2f\left(t + \frac{x}{C_0}\right) \qquad \text{(Equation 2.27)}$$

Experience and Equation (A2.6) show that a decrease in velocity is accompanied by an increase in pressure head. Hence, a wave of decreasing velocity in the direction of negative velocity must be accompanied by an increase in pressure head which can only be satisfied by Equation (2.27). Whatever convention is adopted for the sign of x, which then determines the argument for the arbitrary functions, the choice of Equation (2.26) or (2.27) is determined only by the convention for positive velocity. Thus if v is positive downstream, Equation (2.26) would propagate downstream and (2.27) would propagate upstream.

Appendix 3

Basic method of characteristics

The numerical solution of water hammer is most easily achieved by the digital computer using equations which have been expressed in finite difference form. The simultaneous partial differential equations are first transformed into ordinary differential equations (Gray, 1953):

$$\lambda\frac{\mathrm{d}h}{\mathrm{d}t}+\frac{\mathrm{d}v}{\mathrm{d}t}+\frac{f}{2D}v|v| = 0 \tag{A3.1}$$

with

$$\lambda = \pm g/C_0 \tag{A3.2}$$

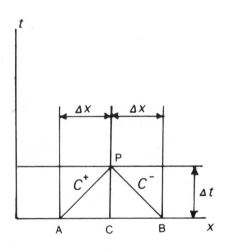

Fig. A3.1

Numerical solutions by the method of characteristics treat the problem in the x, t domain, as opposed to the h, v domain of the graphical method. In the x, t plane solutions are valid along characteristic lines C^+ and C^- (see Fig. A3.1) where $\lambda = +g/C_0$ corresponds to $\mathrm{d}x/\mathrm{d}t = +C_0$ and $\lambda = -g/C_0$ corresponds to $\mathrm{d}x/\mathrm{d}t = -C_0$. The local fluid velocity is assumed to be

131

much less than C_0 which, when constant, will yield straight lines in the x, t plane. In finite difference form the equations become:

$$v_P - v_A + \frac{g}{C_0}(h_P - h_A) + \frac{f}{2D} v_A |v_A| \Delta t = 0 \tag{A3.3}$$

$$v_P - v_B - \frac{g}{C_0}(h_P - h_B) + \frac{f}{2D} v_B |v_B| \Delta t = 0 \tag{A3.4}$$

In Fig. A3.1 the conditions at P at time $t + \Delta t$ can be found from conditions at A and B (when $2\Delta x$ apart), at time t. Equation (A3.3) is valid along a C^+ and Equation (A3.4) along a C^- characteristic.

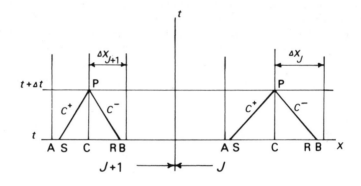

Fig. A3.2

There are a number of different schemes to effect the solution but one will suffice to support this text. With a constant time increment and with different pipes the possibility of different size elements (Δx) must be provided for, as well as the need to have an integer number of partitions in each pipe. Thus in Fig. A3.2, AC in pipe J will not necessarily equal AC in pipe $J+1$, and with integerization, propagations to P along C^+ and C^- characteristics would generally come from points R and S (inside AB). Conditions at R and S can be found by linear interpolation from A and B. For each pipe, therefore, an interpolation ratio must be calculated as follows:

$$\frac{v_C - v_R}{v_C - v_B} = \frac{x_C - x_R}{x_C - x_B} \tag{A3.5}$$

Since $x_p = x_c$ and $C_0 = (x_p - x_R)/\Delta t$

$$v_R = v_C - C_0 R_J (v_C - v_B), \quad h_R = h_C - C_0 R_J (h_C - h_B) \tag{A3.6}$$

and

$$v_S = v_C - C_0 R_J (v_C - v_A), \quad h_S = h_C - C_0 R_J (h_C - h_A) \tag{A3.7}$$

where $R_J = \Delta t/\Delta x_J$ and similarly for pipe $J+1$ where $R_{J+1} = \Delta t/\Delta x_{J+1}$.

Thus in the computer program the first calculations involve finding a value $AN(J) = PL(J)/(DT * CO(J))$ where DT is the time corresponding to the smallest value of $PL(J)/(CO(J) * N)$ for all pipes where N is an integer not less than 2.

$$R(J) = AN(J) * DT/PL(J)$$

and then each pipe is divided into a whole number of elements equal to the least integer value of each $AN(J)$. This scheme ensures that R and S always lie inside A and B, and the solution is stable.

By addition and subtraction of equations (A3.3) and (A3.4) the interior point P values would be determined by the pair of equations:

$$v_P = \frac{1}{2}\left(v_R + v_S + \frac{g}{C_0}(h_R - h_S) - \frac{f}{2D}\Delta t(v_R|v_R| + v_S|v_S|)\right) \tag{A3.8}$$

$$h_P = \frac{1}{2}\left(h_R + h_S + \frac{C_0}{g}(v_R - v_S) - \frac{C_0}{g}\frac{f}{2D}\Delta t(v_R|v_R| - v_S|v_S|)\right) \tag{A3.9}$$

Solution equations

Listed briefly below are the finite difference equations for other cases that are illustrated in the text.

Upstream reservoir

For an upstream reservoir only one equation of type (A3.4) is necessary to find v_p, since h_p is constant (for an infinite reservoir) and thus:

$$v_P = v_R + \frac{g}{C_0}(h_p - h_R) - \frac{g}{C_0}F_K v_R|v_R| \tag{A3.10}$$

where $\frac{f}{2D}\Delta t v^2$ has been written as $\frac{g}{C_0}F_K v^2$ for the programming examples.

Downstream gate valve

In this case at the gate

$$v_p = v_S - \frac{g}{C_0}(h_p - h_S) - \frac{g}{C_0}F_K v_S|v_S| \tag{A3.11}$$

which applies simultaneously with the equation describing the gate characteristic (see Equation (11.2)):

$$v_p{}^2 = \beta^2 \frac{v_0{}^2}{ph_0}h_p \tag{A3.12}$$

where ph_0 is the pressure head lost across the gate when $\beta = 1$, i.e. fully open.

Equation (A3.11) may be written as:

$$v_P = v_N - \frac{g}{C_0} h_P \tag{A3.13}$$

where

$$v_N = v_S + \frac{g}{C_0} h_S - \frac{g}{C_0} F_K |v_S| v_S \tag{A3.14}$$

In Equation (A3.14), v_N expresses conditions near the gate in the previous time interval and hence it is known. Combining Equations (A3.12) and (A3.13):

$$v_P = -\frac{v_0{}^2}{ph_0} \frac{C_0}{2g} \beta^2 \left(1 - \sqrt{1 + \frac{4 v_N g \, ph_0}{v_0{}^2 C_0 \beta^2}}\right) \tag{A3.15}$$

Thus, given β, the value of v_P (and also h_P) can be found. When $\beta = 0$, i.e. gate fully closed, Equation (A3.15) cannot be used and an alternative calculation with $v_P = 0$ is inserted in the computer program.

Pipes in series

At a junction, continuity of flow must be satisfied and the pressure head at any instant must be same in each pipe. If the point P is at the junction, then the upstream side may be designated P_S and the downstream side P_R. Hence:

$$v_{P_R} A_R = v_{P_S} A_S \text{ and } h_{P_R} = h_{P_S}$$

In a manner similar to Equation (A3.13), expressions may be written for the junction in terms of previous conditions from each side as follows:

$$v_{P_R} = v_R + \frac{g}{C_{0_1}} (h_P - h_R) = v_{RR} + \frac{g}{C_{0_1}} h_P \tag{A3.16}$$

$$v_{P_S} = v_S - \frac{g}{C_{0_2}} (h_P - h_S) = v_{SS} - \frac{g}{C_{0_2}} h_P \tag{A3.17}$$

With the compatibility conditions described above, it can be shown:

$$h_P = \frac{-v_{RR} A_1 + v_{SS} A_2}{\dfrac{g}{C_{0_1}} A_1 + \dfrac{g}{C_{0_2}} A_2} \tag{A3.18}$$

The value of the velocity in each pipe at the junction is then readily calculated.

Pipe junction – tee

At a junction of three pipes the continuity equation will be (referring to Fig. A3.3):

$$v_P'' A_2 = v_P' A_1 + v_P''' A_3 \tag{A3.19}$$

and the value of the pressure head must be the same in each pipe at the junction at all times. There will be three simultaneous equations describing conditions in each pipe at the junction as follows:

$$v_P' - v_R = \frac{g}{C_{0_1}} (h_P - h_R) \tag{A3.20}$$

$$A_1 v_P' = \frac{g}{C_{0_1}} A_1 h_P + v_{RR} A_1 \tag{A3.20a}$$

$$v_P'' - v_S = -\frac{g}{C_{0_2}} (h_P - h_S) \tag{A3.21}$$

$$A_2 v_P'' = -\frac{g}{C_{0_2}} A_2 h_P + v_{SS} A_2 \tag{A3.21a}$$

$$v_P''' - v_{IR} = \frac{g}{C_{0_3}} (h_P - h_{IR}) \tag{A3.22}$$

$$A_3 v_P'' = \frac{g}{C_{0_3}} A_3 h_P + v_{IRR} A_3 \tag{A3.22a}$$

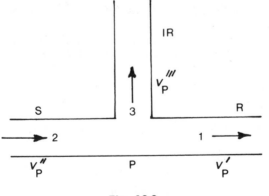

Fig. A3.3

In the Equations (a) above, the terms written as v_{RR}, v_{SS} and v_{IRR} respectively incorporate information known from the previous time interval. Combining these with the compatibility conditions at the junction the resultant pressure head may be expressed as:

$$h_P = \frac{-v_{RR} A_1 + v_{SS} A_2 - v_{IRR} A_3}{\dfrac{g}{C_{0_1}} A_1 + \dfrac{g}{C_{0_2}} A_2 + \dfrac{g}{C_{0_3}} A_3} \tag{A3.23}$$

The value of the velocity in each pipe may now be calculated from the Equations (a) above.

Rupture of the water column

Recently, investigators have applied the method of characteristics to the very complex problem of rupture of the water column (Safwat, 1972; Baltzer, 1967; Kalkwijk and Kranenburg, 1971). Quite complicated interactions must be anticipated at the interface between the liquid and the gas (and vapour) phases, and in this introductory text no attempt will be made to set up a combined (liquid and gas) solution. A suitable simulation of rupture for a gradually rising pipeline is possible for the situation where power failure to a pump occurs (Sharp, 1977). To enable these calculations to be made it was assumed that cavities would occur at positions spaced equally apart on the rising limb and that the cavity growth and collapse was dominated by conditions on the downstream side of the cavity.

Pumps

Solutions for sudden changes in pump operation may be readily obtained by numerical methods provided a scheme for representing the required range of pump data can be developed. Pump data may be made dimensionless and homologous so that the zones of possible operation including the dissipation ranges and running as a turbine can be described by one set of values for a specific speed (Donsky 1961) and, hence, for a class of pump.

Typical pump data

To demonstrate a solution technique for the normal zone of pump operation for a pump with a specific speed of 34.9 (metres), 1800 (US gpm, ft) the data for head (H) and torque (β) are listed below:

V/α	0	0.1	0.2	0.3	0.4	0.5	0.6	0.7	0.8	0.9	1.0
H/α^2	1.288	1.287	1.284	1.272	1.256	1.236	1.206	1.166	1.118	1.061	1.00
β/α^2	0.450	0.504	0.567	0.633	0.707	0.772	0.833	0.886	0.931	0.969	1.00

V/α	1.1	1.2	1.3	1.4	1.5	1.6	1.7	1.8	1.9	2.0
H/α^2	0.913	0.812	0.707	0.599	0.489	0.381	0.271	−0.124	−0.133	−0.320
β/α^2	1.008	1.008	1.006	0.996	0.989	0.981	0.963	−0.931	−0.927	−0.860

The pump data is referred to suction conditions as zero head and as this is often not the case arrangements have to be made in the transfer to and from a pump subprogram to convert to the desired reference suction level.

Changes in pump operation

It is usual to guess or extrapolate suitable values for speed and velocity after the time interval has elapsed. From the typical pump data above, given V/α, new values of H/α^2 and β/α^2 can be found by interpolation and hence values of H and β.

The change in torque is associated with a speed change by the relation

$$\alpha_1 - \alpha_2 = M(\beta_1 + \beta_2)\Delta t \tag{A3.24}$$

where

$$M = 1.79 \times 10^6 \, HQ/(GD^2\eta N^2) \tag{A3.25}$$

Referring to Fig. A3.4, the propagation of previous values v_p, h_p from a point nearby on the delivery side will give new conditions VX, HX at the new speed α_2 and there will be a corresponding new β_2 at this point. All variables are normalized at this stage.

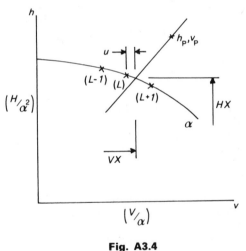

Fig. A3.4

One can use an interpolation formula such as Stirling's for equally spaced tabulated data and the equations are set up as follows. New values of VX and α are first extrapolated and using

$$HX - h_P = \frac{C_0}{g}(VX - v_P) \tag{A3.26}$$

an expression for the difference u (see Fig. A3.4) from the tabulated values is

$$\frac{VX}{\alpha} \cdot 10 - L = u = \frac{g\,10}{\alpha C_0}(HX - h_P) + \frac{10 v_P}{\alpha} - L \qquad (A3.27)$$

where L is the absolute value of $10V/\alpha$.

Hence, if X, Y and Z are the $L-1$, L and $L+1$ values of H/α^2, from the pump data table (p. 136), at the new speed:

$$HX = Y + \frac{u}{2}(Z - X) + \frac{u^2}{2}(Z - ZY + X) \qquad (A3.28)$$

and on combining with Equation (A3.26) a quadratic in HX is obtained. The result provides an HX, VX that satisfies both the propagation requirements and the torque speed changes in the specified time interval. The method depends on suitable extrapolation formulae for speed and velocity.

Formulae which produce satisfactory extrapolation are:

$$\alpha = 3\alpha' - 3\alpha'' + \alpha''' \qquad (A3.29)$$
$$V = V' + (V' - V'')/2 + (V'' - V''')/4 \qquad (A3.30)$$

where the primes indicate the previous three values, one prime being the most recent. This requires the first three values of both speed and velocity to be specified initially. If the values of the pump data near the initial point such as rated head and discharge (for a pump failure case) are connected by a quadratic

$$\frac{H}{\alpha^2} = A\left(\frac{V}{\alpha}\right)^2 + B\frac{V}{\alpha} + C$$

then $dH = 2A\,dV + B\,d\alpha + B\,dV + 2C\,d\alpha$ and $dx = 2K\beta\Delta t$ from the pump torque equation.

Thus the decrement in pump velocity from the rated conditions becomes for the pump data (p. 136):

$$dV = 0.44 d\alpha/(C_0/g + 0.74) \qquad (A3.31)$$

The procedure then is to enter the subprogram with previous values of speed and velocity as well as the most recent value of H and V for the nearby interior point on the delivery side.

Table look-up procedures are used to find the most suitable values from which to interpolate to find the new H for the extrapolated V and α. The conditions at the pump are then known as new boundary conditions which are propagated back into the system.

References

ALLIEVI, L., *The Theory of Water Hammer*, (1913) Translated by E. E. Halmos, Riccardo Gatroni, Rome, Italy, 1925, (ASME).

ANGUS, R. W., Simple graphical solution for pressure rise in pipes and pump discharge lines, *J. Eng. Inst. Canada*, Feb 1935, pp. 72–81.

ANGUS, R. W., Waterhammer pressures in compound and branched pipes, *Proc. ASCE*, No. 2024, Jan 1938, pp. 340–401.

BALTZER, R. A., Column separation accompanying liquid transients in pipes, *J. of Basic Eng. Trans. ASME*, Series D, **89**, 1967, pp. 837–46.

BEATTIE, D. R. H., *Pressure Pulse and Critical Flow Behavior in Distributed Gas-Liquid Systems*, 2nd Int. Conf. on Pressure Surges, London, Sept 1976, paper D3.

BERGERON, L., *Waterhammer in Hydraulics and Wave Surges in Electricity*, John Wiley & Sons, New York, 1961.

BERNHART, H. H., *The Dependence of Pressure Wave Transmission through Surge Tanks on the Valve Closure Time*, 2nd Int. Conf. on Pressure Surges, London, Sept 1976, paper K3.

CABELKA, J. and FRANC, I., *Closure Characteristics of a Valve with respect to Water Hammer*, 8th Cong. of IAHR, Montreal, Canada, 1959, p. 6-A-1.

CROWE, T. J., *The Water Hammer Gate Characteristic*, MEngSc Thesis, U of Melbourne, 1968.

DIJKMAN, H. K. M. and VREUGDENHIL, C. B., The effect of dissolved gas on cavitation in horizontal pipelines, *J. Hyd. Res.*, 7, No. 3, 1969, p. 301.

DONSKY, B., Complete pump characteristics and the effects of specific speeds on hydraulic transients, *J. Basic Eng.*, Dec 1961, pp. 685–99.

DUC, J., *Negative Pressure Phenomenon in Pump Pipelines*, ASME International Symp. on Waterhammer in Pumped Storage Projects, Nov 1965, Chicago, pp. 154–67.

ENEVER, K. J., *Surge Pressures in a Gas–Liquid Mixture with low Gas Content*, Int. Conf. on Pressure Surges, Canterbury, 1972, paper C1.

FAVRE, H., Theorie des coups de belier dans les conduites a caracteristiques lineairement variables le long de l'axe, *Rev. gen. Hydraulique*, Nos. 19, 20, 21, 22, 23 and 24, 1938.

GERNY, J. S., *An investigation into the Cause and Mitigation of Water-Hammer effects in rising mains with particular reference to the Morgan-Whyalla Pipeline*. ME Thesis, University of Adelaide, 1949.

GIBSON, A. H., *Hydraulics and its Applications*, Constable, 4th Edition, 1945, p. 234.

139

GLOVER, E. R., *Computations of Water Hammer Pressures in Compound Pipes*, Symp. on Waterhammer, ASME-ASCE, 1933, pp. 64–69.

GRAY, C. A. M., The Analysis of the Dissipation of Energy in Waterhammer, *Proc. ASCE Paper 274*, **119**, 1953, pp. 1176–94.

GRAZE, H., *A Rational Thermodynamic Equation for Air Chamber Design*, 3rd Australasian Conf. on Hyd. and Fluid Mechs, Sydney, 1968, pp. 57–61.

HARVEY, E. N., McELROY, W. D. and WHITELY, A. H., On cavity formation in water, *J. Appl. Physics*, **18**, 1947, p. 162.

HOLDER, D. W., STUART, B. A. and NORTH, R. J., The interaction of a reflected shock with the contact surface and boundary layer in a shock tube, *Aeronautical Res. Council Rep.*, **22**, 891, *Hyp.*, **191**, Sept 1961.

JAEGER, C., *Engineering Fluid Mechanics*, Blackie and Sons Ltd., 1956.

JAEGER, C., Theory of resonance in hydropower systems. Discussion of incidents and accidents occurring in Pressure Systems, *Trans. ASME, J. of Basic Eng.*, Series D, **85**, No. 3., Sept 1965, pp. 631–40.

JOHNSON, V. E., Cavitation hydraulic structures: mechanics of cavitation, *Proc. ASCE, J. of Hyd. Div.*, **89**, 1963, p. 251.

JOUKOWSKI, N., Waterhammer, translated by O. Simin, *Proc. AWWA*, **24**, 1904, pp. 341–424.

KALKWIJK, J. P. Th. and KRANENBERG, C., Cavitation in horizontal pipelines due to water hammer, *J. of Hyd. Div.*, *Proc. ASCE*, **97**, HY10, Oct 1971, p. 1585.

KINNO, H., *Dynamics of Waterhammer Control Systems*, JSME Semi-international Symposium, Tokyo, Sept 1967, pp. 223–32.

KNAPP, R. T., Complete characteristics of centrifugal pumps and their use in prediction of transient behaviours, *Trans. ASME*, **59**, 1937, pp. 683–689.

KRANENBURG, C., *The effect of Free Gas on Cavitation in Pipelines*, 1st Int. Conf. on Pressure Surges, Canterbury, Sept 1972, paper C4.

KRANENBURG, C., Gas release during transient cavitation in pipes, *Proc. ASCE, J. Hyd. Div.*, **100**, No. 10, Oct 1974, p. 1383.

LI, W. H., Mechanics of Pipe-Flow following Column Separation, *J. of Eng. Mechs., ASCE*, **88**, No. EM4, 1962, p. 97.

LINTON, P., *Notes on Pressure Surge Calculations by the Graphical Method*, BHRA publication TN 447, May 1954.

LUPTON, H. R., Graphical analysis of pressure surges in pumping systems, *J. Instn. of Water Engs*, **7**, 1953, p. 87.

MACLELLAN, D. A. S., and CARRUTHERS, J. H., *Hydraulic Characteristics ana Limitations of Butterfly Valves for Flow Control*, Valve Symposium – Fluid Control in Industry, Earls Court, London, May, 1976.

MARTIN, C. S., PADMANABHAN, M. and WIGGERT, D. C., *Pressure Wave Propagation in Two-Phase Bubbly Air-Water Mixtures*, 2nd Int. Conf. on Pressure Surges, London, Sept 1976, paper C1.

MICHEL, B., Discussion (see Cabelka and Franc, 1959).

MIYASHIRO, H., Water Hammer Analysis of Pump Discharge line with Several one-way Surge Tanks, *Trans. ASME, J. Eng. for Power*, Oct 1967, pp. 621–7.

PARMAKIAN, J., *Water Hammer Analysis*, Dover Publications, New York, 1963.

PAYNTER, H. M., and EZEKIEL, F. D., Water hammer in non-uniform pipes as an example of wave propagation in gradually varying media, *Trans. ASME*, Paper No. 57-A-107, 1958, pp. 1585–95.

SAFWAT, H. H., Photographic study of water column separation, *J. of Hyd. Div., Proc. ASCE*, **98**, No. HY4, April 1972, pp. 739–46.

SAFWAT, H. H., Measurement of transient flow velocities for water hammer applications, *Trans. ASME, J. Fluids Eng.*, **29**, 1972.

SHARP, B. B., *The Growth and Collapse of Cavities produced by a Rarefaction Wave with Particular Reference to Rupture of the Water Column*, PhD Thesis, U of Melbourne, March 1965.

SHARP, B. B., Blast wave propagation in hydraulic conduits, Discussion of paper by P. Lieberman, *Trans. ASME., J. Eng. for Power*, Oct 1965, p. 439.

SHARP, B. B., *Rupture of the Water Column*, 2nd Australasian Conf. on Hyd. and Fluid Mechs., Auckland, NZ, (Dec 1965), 1966, pp. A169–76.

SHARP, B. B., The water hammer gate characteristic, *Water Power*, Sept 1969, p. 352.

SHARP, B. B., Discussion to paper by Wood and Jones (1973), *Proc. ASCE, J. Hyd. Div.*, **100**, Hy2, Feb 1974, p. 323.

SHARP, B. B., *Water Hammer in Australia (a review)*, Proc. Thermofluids Conf. IEA, Melbourne, Dec 1974.

SHARP, B. B., A simple model for water column rupture, *Proc. 17th Cong. IAHR*, Baden Baden, West Germany, 1977.

SHARP, B. B. and COULSON, G. T., Surge problems in the water supply mains to Hazelwood Power Station, *J. of IEA*, Australia, 1968, p. 83.

SIEMONS, J., The phenomenon of cavitation in a horizontal pipeline due to a sudden pump failure, *J. of Hyd. Res.*, **5**, No. 2, Feb 1967, p. 135.

STEPHENSON, D., *Discharge Tanks for Suppressing Water Hammer in Pumping Lines*, Int. Conf. on Pressure Surges, Canterbury, 1972, paper F3.

STREETER, V. L., *Water Hammer Analysis of Pipelines*, Univ. of Mich., Ann Arbour, Nov 1963.

STREETER, V. L. and WYLIE, E. B., *Hydraulic Transients*, McGraw Hill, 1967.

SWAFFIELD, J. A., A study of the influence of air release on column separation in an aviation kerosine pipeline, *Proc. Inst. Mech. Engs*, **186**, 1972, p. 693. *Symposium on Water Hammer*, ASME–ASCE, 1933.

THORLEY, A. R. D., Modern methods of analysing resonance in hydraulic systems, *Water Power*, July 1973, pp. 250–3.

TULLIS, T. P., STREETER, V. L., and WYLIE, E. B., *Water Hammer Analysis with Air Release*, 2nd Int. Conf. on Pressure Surges, London, Sept 1976, paper C3.

VAN EMMERIK, F. L., *The Attenuation of Plane Shock Waves in a Liquid*, MEngSc Thesis, U of M, 1964.

WALLER, E. J., *Prediction of Pressure Surges in Pipelines by Theoretical and Experimental methods*, Oklohoma State University, Pub. 101, Stillwater, Oklahoma, June 1958.

WALLER, E. J., *Pressure Surge control in Pipeline Systems*, Oklohoma State University, Pub. 102, Stillwater, Oklahoma, Jan 1959.

WOOD, F. M., Computer solutions of surges in a simple inclined pipe line for cases of high friction losses and low wave velocity, *Trans. Eng. Inst. of Canada*, Dec 1966.

WOOD, D. J. and JONES, S. E., Water hammer charts for various types of valves, *Proc. ASCE, J. Hyd. Div.*, 1973.

WOOD, D. J. and STELSON, T. E., Energy Analysis of Pressure Surges in Closed Conduits, *Experimental and Applied Mechanics*, 1966, pp. 371–88.

WYLIE, E. B., Resonance in pressurized piping systems, *Trans. ASME, J. of Basic Eng.*, **87**, Series D, No. 4, Dec 1965, pp. 960–6.

ZIELKE, W., Frequency dependent friction in transient pipe flow, *Trans. ASME, J. of Basic Eng.*, March 1968, pp. 109–15.

ZIELKE, W. and HACK, H. P., *Resonance Frequencies and associated Mode Shapes of Pressurised Piping Systems*, Int. Conf. on Pressure Surges, Canterbury, 1972, paper G1.

General references

Symposium on Water Hammer, Joint Committee of ASME and ASCE, New York, 1933.

Proc. 1st Int. Conf. on Pressure Surges, Canterbury, 1972.

Proc. 2nd Int. Conf. on Pressure Surges, London, 1976.

WOOD, F. M., *History of Water Hammer*, Queen's University at Kingston, C E Res. Rep., No. 65, April 1970.

MARTIN, C. S., Status of fluid transients in Western Europe and the United Kingdom, Report on Laboratory Visits by Freeman Scholar, *Trans. ASME, J. of Fluids Eng.*, June 1973, pp. 301–18 and discussions.

Index